Cloud Security

Cloud Security

Attacks, Techniques, Tools, and Challenges

Preeti Mishra
Doon University, Dehradun

Emmanuel S Pilli
Malaviya National Institute of Technology, Jaipur

R C Joshi
Graphic Era Deemed to be University, Dehradun

CRC Press is an imprint of the
Taylor & Francis Group, an **informa** business
A CHAPMAN & HALL BOOK

First edition published 2022
by CRC Press
6000 Broken Sound Parkway NW, Suite 300, Boca Raton, FL 33487-2742

and by CRC Press
2 Park Square, Milton Park, Abingdon, Oxon, OX14 4RN

ISBN: 978-0-367-43582-0 (hbk)
ISBN: 978-1-032-19026-6 (pbk)
ISBN: 978-1-003-00448-6 (ebk)

DOI: 10.1201/9781003004486

Typeset in LM Roman
by KnowledgeWorks Global Ltd.

To my mother Mrs. Sarla Mishra, father Mr. Diwakar Mishra, husband Mr. Deepak Joshi, daughter Divyanshi Joshi, father-in-law Mr. P. C. Joshi and mother-in-law Mrs. Kamla Joshi and PhD supervisors Dr. Emmanuel S. Pilli and Prof. Vijay Varadharajan-P. M

To my brothers Daniel Charles Pilli and Samuel Sudhakar Pilli and their families—E. S. P

To my brother Late Shri A. B. Joshi, my wife Smt Usha Joshi, daughters Ira and Bakul —R. C. J

and all faculty members and students who are working hard to make the cyberworld a safe place

Contents

III Tools and Advances 131

Preface

We are living in the era of cloud computing, where services are provisioned to the users on demand and 'pay-per-use' basis from a resource pool. Cloud computing has evolved gradually over a period of time. National Institute of Standards and Technology (NIST) defines cloud computing as a model for enabling convenient, ubiquitous and on-demand network access to a shared pool of computing resources (e.g., servers, network, storage and applications) that can be rapidly provisioned and released with minimal management effort or service provider interaction". Users are gradually adopting cloud services because of the ease and flexibility with cloud services. Most of the companies are changing the way they operate and moving toward cloud-based services.

However, attacking incidents are also increasing day by day with the evolution of cloud computing. Security in such a complex technological environment is very important for providing assurance to cloud customers. Any vulnerability present in cloud, can allow the attacker to gain illegal privileges of Virtual Machine (VM) users. A malicious user can install advanced malware programs and gain higher access privileges (guest OS kernel privilege). A compromised guest kernel can call malicious drivers and can perform malicious actions. Once a VM is fully compromised, an attacker can try to launch attacks such as spreading malwares (virus, worm, etc.), flooding and scanning other VMs. A compromised VM is a big threat to cloud infrastructure which can bypass the security of other VMs. It could further lead to monetary disputes between cloud service provider (CSP) and legitimate VM users. Other than VM Security, there exist various other security issues related to application level, Network level, Virtualization level, Data storage level, Identity management and Role-based access control, Cryptographic key management level, SLA and trust level, Auditing, governance and regulatory compliance and Cloud & CSP migration level security, discussed in the book in detail.

Hence, the importance of well-organized architecture and security roles have become greater with the popularity of cloud computing. People are working in the cloud security domain have proposed various security frameworks to tackle with security threats. The existing frameworks that deploy security tool at individual Tenant Virtual Machine (TVM) are prone to subversion attacks. They are less efficient in detecting malicious activities. Moreover, the TVM-layer security solutions cannot be directly applied at the Virtual Machine Monitor (VMM)-layer because of the semantic gap problem at the hypervisor. Semantic gap refers to interpreting the low-level information of a guest OS into a high-level semantics. VM introspection (VMI) is one of the virtualization-specific approaches that provides possible ways to obtain the high-level view of TVM at hypervisor.

However, not enough work has been done in this direction to provide VMI-based security solutions for cloud. The existing VMM-layer solutions do not provide a complete solution to detect both basic and evasive malware attacks in cloud. On the other hand, some of the cloud security frameworks are designed to detect network intrusions only. Most of them apply signature-matching technique as core detection technique, making them prone to signature-manipulation attacks. This book provides an in depth understanding various security techniques with their short comings. It also talks about various advances in cloud security.

In this book, we have endeavored to provide a technical foundation that will be practically useful not just for professional cloud security analysts conducting security practices but also for students, independent researchers, and all those who are curious in the field of cloud security.

Audience:

This book is intended for both academic and professional audiences. As a textbook, it is intended as a semester course at under graduate and post graduate level students in Computer Science, Information Technology, Network Security, and Information Science and Management. The book serves as basic reference volume for researchers in cloud security. It will be useful to practitioners, cloud security team, and the cloud security auditor as well. To get the most out of this book, the reader should have a working knowledge of various operating system environments, hypervisors, cloud computing fundamentals, programming languages like Python and Java, and a working knowledge of security tools.

Organization of the book:

The book is organized to provide a broad overview of the important topics of Cloud security. It is divided into three parts: "Fundamentals," "Threat model, attacks & defensive techniques," and "Tools & Advances."

Part I, "Fundamentals," covers the basic concept of cloud computing and Cloud Security. This provides a foundation for more advanced topics, which are covered in the next two parts. Part I includes the following chapters:
Chapter 1: "Introduction to Cloud Computing" presents an introduction to key domain considered and gives the brief background history of Cloud Computing. The chapter also discusses the characteristics, service models and deployment models and associated open research challenges.
Chapter 2: "Introduction to Cloud Security" presents various vulnerabilities along with cloud security concepts, standards and cloud security reference architectures.

Chapter 3: "Cloud Security and Privacy Issues" presents various cloud security goals and concepts, security issues, requirements for privacy and security.

Part II, "Threat model, attacks, defensive systems and security techniques," discusses threats and attacks along with major mechanisms which can be applied to cloud security.

Chapter 4: "Threat model and Cloud Attacks" covers the threat model and various possible attacks at various layers in Cloud Computing.
Chapter 5: "Classification of various IDS in Cloud" covers the types and characteristics of various Cloud-IDS and provides future research directions.
Chapter 6: "Intrusion Detection Techniques in Cloud" discusses various misuse, anomaly, virtual machine introspection and hypervisor introspection techniques used to protect the cloud from attacks.

Part III, "Tools and Advances," covers various tools and advance topics such as introspection and container security.

Chapter 7: "Overview of Tools in Cloud" covers the classification of various attacking and security tools and case study of LibVMI, a hypervisor-based security tool.
Chapter 8: "Virtual Machine Introspection and Hypervisor Introspection" covers the advanced virtualization specific cloud security techniques used to protect the virtual domain and hypervisor in Cloud.
Chapter 9: "Container Security" covers the threat model and attacks in containerized environment. It also discusses various defensive mechanisms and open challenges. A case study on the Sql Injection attack in Docker systems is also demonstrated.

Tools:

This book is designed to be accessible to a wide audience to teach the fundamental principles and techniques of cloud security. There are many tools available to perform various attacking activities, taking memory snapshots both from inside the VM and outside the VM and analyzing and extracted log files. The focus is to provide the technical insight by providing the detailed classification of attacking and security tools along with case studies of some attacking and security tools.

Acknowledgment

This book is a result of our research work at the Malaviya National Institute of Technology Jaipur, Graphic Era University Dehradun and Doon University Dehradun. The book has been inspired by our research works published in reputed journals in the area of Cloud Security. I would like to offer special thanks to co-authors Prof. Vijay Varadharajan (Global Innovative Chair in Cybersecurtiy, The University of Newcastle Callaghan, Australia) and Dr. Udaya Tupakula (Senior Lecturer, The University of Newcastle Callaghan, Australia) for their excellent guidance. We have thoroughly updated our work and included state-of-the-art practices, reference architectures, standards, security and attack tools, and case studies in Cloud Security. We have included advanced topics such as Virtual Machine Introspection, Hypervisor Introspection, and Container Security.

We would like to thank Prof. Surekha Dangwal, Vice Chancellor Doon University Dehradun, Prof. Udaykumar R. Yaragatti, Director MNIT Jaipur and Prof. Kamal Ghanshala, President Graphic Era Deemed to be University Dehradun who have wholeheartedly supported the writing of this book.

We gratefully recognize the opportunity Taylor and Francis gave us to write this book. We like to make a special mention of thanks to all the splendid staff specially Shikha Garg (Senior Editorial Assistant, CRC Press—Taylor & Francis Group), Aastha Sharma (Senior Acquisitions Editor, CRC Press—Taylor & Francis Group), Isha Singh (Editorial Assistant, CRC Press- Taylor & Francis Group), and Shashi Kumar (Senior Assistant Manager, KnowledgeWorks Global Ltd) who put so much time and effort into producing this book. They were ever ready to incorporate many editing changes made by us during the proof reading phase.

Dr. Mishra would like to thank her entire family as the journey would never have been possible without their support. She specially thanks her mother Mrs. Sarla Mishra and father Mr. Diwakar Mishra for always supporting her in all possible ways in all difficult and good times. She also thanks her husband Mr. Deepak Joshi for his support and motivation, daughter Divyanshi Joshi for keeping her active and happy during this period. She also thanks her father-in-law Mr. P. C. Joshi and mother-in-law Mrs. Kamla Joshi, for always motivating her to do something good. She thanks her brother Mohit and sister-in-law Pinky, for having their joyful company during the final editing phase of this book. Special thanks to her friends Mini Kandpal, Reeta Uniyal, Asha,

Amita, Jharna, Nisha, Sonal, Ritu, Gaurav Varshney, and Ankit Vidyarthi for giving a wonderful company and helping her to overcome stress.

She would like to specially thank the research scholars Umang Garg, Divya Kapil and under graduate students Saurabh Gupta and Phalugni, Saloni for their help during the work and for being the part of her research team. She thanks to the other research team members as well specially under graduate students Palak, Ishita, Kashish, Akansh, Sachin, Shivam, Garima, Rahul Bisht, Rahul Sharma and post graduate students Aparna and Diksha and research scholars Sarishma, Charu Negi for being the part of her security group, named CyberZine.

She would also like to thank her foreign research collaborators specially Dr. Nour Moustafa (Senior Lecturer in Cyber Security and Computing at the School of Engineering and Information Technology (SEIT), University of New South Wales (UNSW)'s UNSW Canberra, Australia) and Dr. Zakirul Alam Bhuiyan (Assistant Professor, Department of Computer and Information Sciences, Director Dependable and Secure System Research (DependSys), Fordham University USA) who are working in the field of security and privacy and have always inspired her to work hard.

Dr. Emmanuel Pilli would like to thank his wife Phoebe Vanmathy Julius and their daughter Pramiti Evangeline.

Prof. Joshi would like to thank all his family members, especially his wife Smt. Usha Joshi and daughters Ira and Bakul.

List of Figures

List of Tables

Author Bios

Preeti Mishra is currently working as Assistant Professor in the Department of Computer Science, Doon University, Dehradun, India. Earlier, she was associated with Graphic Era Deemed to be University Dehradun. She has 10+ years teaching and research experience. She has been awarded Ph.D. in Computer Science and Engineering from Malaviya National Institute of Technology Jaipur, India, under the supervision of Dr. Emmanuel S. Pilli and Prof. Vijay Varadharajan (2017). She is a B. Tech and M. Tech Gold Medalist. She has published various SCI/SCIE indexed reputed international journals and reputed conference papers in the area of security and privacy. Some of her key research publications have been published in IEEE Transaction on Cloud Computing (with IF 5.720), IEEE Communication Surveys and Tutorials (with IF 23.7), IEEE Transactions on Industrial Informatics (IF: 9+), etc. as main author. She has also published several publications in reputed international conferences. She worked as a visiting scholar in Macquarie University Sydeny under Prof. Vijay Varadharajan in 2015 and has been awarded a fellowship, administered by the Department Administrators in Department of Computing, Macquarie University, Sydney. She has also been awarded by Graphic Era Deemed to be University Dehradun for outstanding contribution in research. Her research proposal valued more than 20 lakhs got approved by SERB-DST, Govt. of India in the area of Cloud Security. Her area of interest includes Cloud Security, E-mail Security and Network Security, Internet of Things, Blockchain, Cyber Security, Mobile Security, Adversarial Machine Learning, etc. She is an active reviewer of many reputed international journals/conferences such as *IEEE Transaction on Network and Service Management, Future Generation Computer Systems, IEEE Journal of Information Security and Applications*, etc. She is currently serving as Lead Guest Editor in IEEE Transaction on Industrial Informatics (TII).

Emmanuel S. Pilli received his Ph.D. from IIT, Roorkee (2012) and is currently Associate Professor, Dept. of CSE in Malaviya National Institute of Technology, Jaipur, India. Pilli Emmanuel Shubhakar has 21 years of teaching, research, and administrative experience. He completed a research project "Investigating the Source of Spoofed E- mails" from UCOST, Dehradun in 2016. He has coauthored a book "Fundamentals of Network Forensics—A Research Perspective" for Springer in 2016. A total of four students have been awarded Ph.D. under his supervision and 12 Ph.D students are pursuing their research. He is Senior Member of both IEEE and ACM. His areas of interest include Security and Forensics, Cloud Computing, Big Data, IoT, Darkweb, and Blockchain, etc. He is member of Cloud Computing Innovation Council of India (CCICI) and Forensic Science Workgroup on Cloud Computing of the NIST, USA.

Dr. R.C. Joshi, former Prof. E. and C.E. Department at IIT Roorkee and Chancellor at Graphic Era University Dehradun, received his B.E. degree from NIT Allahabad in 1967, M.E. 1st Div. with Honors and Ph.D from Roorkee University, now IIT Roorkee, in 1970 and 1980, respectively. He worked as Lecturer in J.K. Institute, Allahabad University during 1967–1968. He joined Roorkee University in 1970 as Lecturer, became Reader in 1980 and Prof. in 1987. He had been Head of Electronics & Computer Engineering from Jan. 1991–1994 and Jan. 1997 to Dec. 1999. He was also the Head of Institute Computer Centre, IIT Roorkee from March 1994 to Dec. 2005. He was on short visiting Professor's Assignment in University of Cincinnati, USA. University of Minnesota, UA and Macquarie University Sydney Australia also visited France under Indo-France collaboration program during June 78 to Nov. 79. Dr. Joshi has guided 27 Ph.Ds, 250 M.Tech, Dissertation, 75 B.E. Projects. He had taught more than 25 subjects in Computer Engineering, Electronics Engineering and Information Technology. He has worked as Principal Investigator in a number of Sponsored Projects of Ministry of Information & Communication Technology, DRDO, AICTE, UNDP, ISEA, etc

Part I

Fundamentals: Cloud Computing and Security

Chapter 1

Overview of Cloud Computing

1.1 Introduction

The era of cloud is the latest trend which provides various types of on-demand services to the user on the basis of 'pay-per-use' manner, depending upon the requirements of the end-users. The cloud computing has gained lots of popularity and gradually it is expanding its services to address millions of user's demand. The vision of cloud computing industries for the 21st century is to grant computing services in a convenient way just like any other basic services like water or electricity [1]. There is no longer a need to invest on IT infrastructure or developing buildings for initial set-up, and hiring skilled workforce, to run a business. Cloud computing allows small business owners to start the business quickly by using cloud services, without thinking about purchasing and setting-ups large infrastructure. Cloud computing uses various technologies such as virtualization, distributed computing, cluster computing, and service-oriented architecture (SOA), etc. [2]. Cloud computing is developing and constantly improving technology that still does not have any unanimous definition. Various opportunities are provided by the cloud computing to the IT industries by offering a variety of services.

The traditional IT enterprise set-up requires a large infrastructure such as big land space, hardware devices, expensive software licenses, and a big team of IT experts for the establishment of the company. As time passes, there is a requirement to upgrade the whole system of hardware and software to maintain the growth and scalability of the company. Hence, it requires lots of money, resources, and time to maintain and provision the services in traditional way. It makes traditional computing less economical way to start a new business or upgrade the existing ones. Therefore, cloud offers a better economical solution to address the need of organizations. In cloud computing, there is no need to care about the failure and maintenance of any hardware and software services. Developers can focus more on their coding skills rather than focusing on setting up the test environment by downloading and installing various software [3].

Cloud computing is one of the evolving technologies which has been widely used for IT outsourcing, infrastructure provisioning, platform provisioning, software provisioning and database provisioning, etc. Let us now define the

DOI: 10.1201/9781003004486-1

cloud computing term in a more formal way. Cloud is a term that has been used historically by the telecommunication industries as an abstraction of the network for the representation of the system diagram. Cloud computing refers to an Internet-centric computing with virtual infrastructure. The National Institute of Standard and Technology (NIST) proposed a definition of cloud computing [4]. As per this definition "Cloud computing is a model for enabling ubiquitous, convenient, on-demand network access to a shared pool of configurable computing resources(e.g., networks, servers, storage, applications, and services) that can be rapidly provisioned and released with minimal management effort or service provider interaction". The definition is more focused on three main characteristics: (i) Cloud services are scalable, (ii) The overall cost is charged on the basis of usage, and (iii) The quality of services is distributed and managed to the clients.

Foster et al. [5] described grid computing along the concept of cloud computing. Authors described the cloud computing as a kind of distributed computing that is focused on the large-scale paradigm that contains a abstracted pool, scalable, virtualized, dynamic, managed computation resources, large storage and virtual platforms, which are delivered on-demand through the Internet. The adoption of cloud services such as deployment environment, infrastructure, or applications has different impact on the industries. There are several perspectives which can have the potential benefits of cloud services such as: (i) It provides a simple process to establish an environment for application development, (ii) It provides the potential to shorten the process of idea to product, (iii) It provides better solution for the business community, (iv) Simplifies the process for application development, and (v) Provides assurance of the quality of services like security and availability when required [6]. However, there are still many issues which need to be addressed by the vendors before provisioning any of the services online.

Let us now understand the technical terminologies of cloud architecture. An OpenStack [7] cloud architecture is considered here a base model. OpenStack is a global leading cloud management software opted by many companies for developing cloud platform for public, private, or hybrid cloud. It will be discussed in forthcoming sections. The key technology in the cloud environment is virtualization which creates an abstraction layer above the underlying hardware or software. It hides the complexity of physical hardware and allows multiple operating systems to run on the same physical machine. The abstraction layer is called as Virtual Machine Monitor (VMM) or Hypervisor. The cloud architecture with Xen as VMM is considered here. Xen VMM is booted first as a primary boot system. Afterward, Linux kernel is loaded as Dom0 domain by the Hypervisor. Dom0 is the privileged domain (administrative VM) which is used to control, configure, and manage all the other VMs by the cloud administrator. Dom0 runs the device drivers and can access the actual hardware, as shown in Figure 1.1. The networking between the TVMs is provided by VMM. Networking in VMM bridges the virtual adapter to the physical adapter. The tenant virtual machines (TVMs) are loaded after

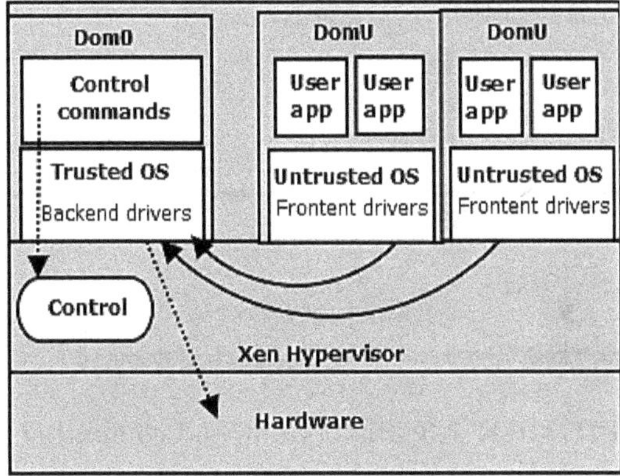

FIGURE 1.1: Basic architecture of Xen.

Dom0 and are also referred as untrusted domains (DomUs). VMM has the highest privilege and full control over any VM running over it. Let us now start understanding the cloud computing architecture.

A cloud environment typically consists of three types of servers: Cloud Controller Server (CCS), Cloud Compute Server (CCoS), and Cloud Networking Server (CNS) [8], as shown in Figure 1.2. The CCS is mainly used to handle all management-related work. The user VMs are hosted in CCoS server. The CNS manages the network, routes the packets, and allocates IPs to the nodes, etc. There are three types of cloud network: administrative, ex-

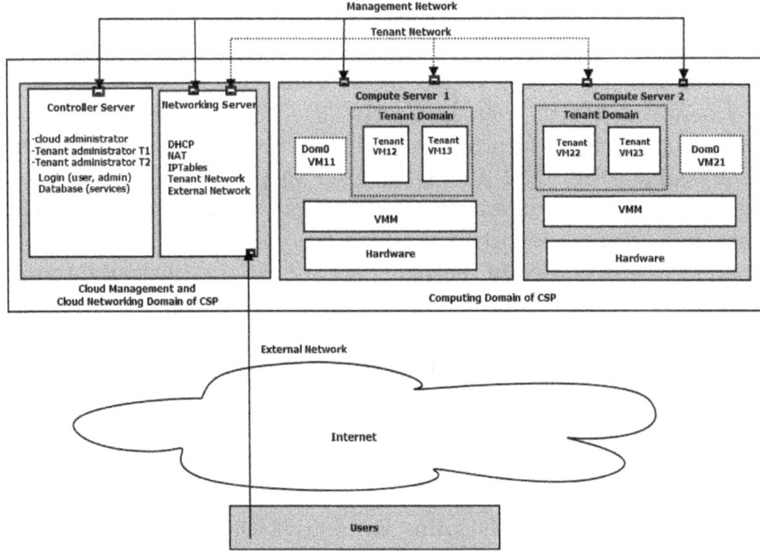

FIGURE 1.2: Basic architecture of cloud environment.

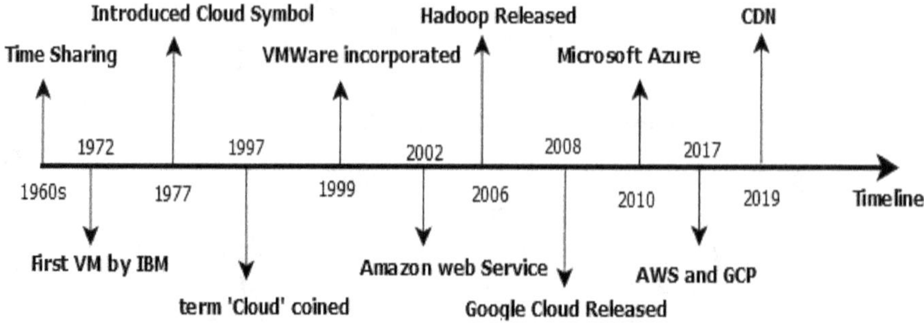

FIGURE 1.3: History of cloud computing.

ternal, and tenant network. The tenant network carries the tenant's data and ensures the end-to-end transportation. Each tenant network connects a set of VMs and is vulnerable to the threats. The administrative network mainly deals with carrying the data corresponding to management commands such as allocating, destroying, creating, and resuming TVM. The external network connects cloud VMs to the external users via Internet. The administrator of the cloud can configure the entire access control policies and has got highest privileges in cloud environment.

1.2 History and Underlying Technologies

Basically, cloud computing is not a new idea or a new technology. It has been evolved over the years. A brief history of cloud computing is shown in the Figure 1.3. In the 1960s, John McCarthy [9] already anticipated that there will be some computing facilities available for the general public like a utility. In the same year, IBM and DEC used to provide their computers on sharing basis [10]. IBM announced its first VM which runs on the physical hardware, and creates the virtual machine environment in the year 1972 [11]. The cloud symbol was used in the year 1977 [12]. In the 1990s, the term cloud has been used to describe a large ATM network. After the evolution of Internet and world wide web, the term 'cloud' is coined by Ramesh Chellapa in the year 1997 [13]. In the year 2006, it was used to describe a business model by Google's CEO Eric Schmidt [14]. After that, it gained a lot of popularity. Since then, cloud computing has been mainly used to represent as a marketing term in several business ideas. In July 2008, some big giants like HP, Yahoo, and Intel announced a global computing laboratory, cloud computing, which enriched with a variety of platforms for sharing the resources and technology in the area of cloud [14]. Voas and Zhang [15] demonstrated six different phases of cloud computing evolution. Firstly, end-users can interact with a mainframe which supports multiple applications. Users can interact directly with their own system which drops the hardware cost with high computational power. Resource sharing can be done through a type of network either LAN or WAN in the third phase. The Internet services can provide access to remote services with an active account pay-per-basis. The next phase provides high-performance computing and throughput using grid computing. Finally, it provides all computing services through the Internet.

The concept of cloud computing is an outcome of evolution and advancements of many other technologies, which makes cloud computing more powerful. As the technologies grow, it introduces several emergences with other key technologies. However, there are some key technologies that played a key role in the development of cloud computing like mainframe, grid computing, web 2.0, or service-oriented computing, etc. Let us now discuss some of the key technologies:

1.2.1 Mainframe computing

Mainframe [16] is one of the key technologies which has large computational facilities with multiple processing units. It provides large input-output operations, reliable computational technology, massive data movements, and powerful tools. Generally, this kind of technology is used by large industries for the transfer of bulk data, online transactions, resource planning, and operations that are dealing with large input-output operations. Mainframes offer a large computational platform with multiple processors that has the ability to process the data as a single entity. Mainframes have a great feature of sustaining computationally expensive load for many hours. It can handle all kinds of failures transparently. The systems were highly reliable due to the replacement of the faulty components without shutting down the system. The key feature of the mainframe systems was batch processing. Although the popularity of these systems is reduced nowadays. Some latest versions of the system are still used in online transactions, airline ticketing, and several government online platforms, etc.

1.2.2 Cluster computing

Cluster computing [17] is an alternative or a low-cost choice to the supercomputers or mainframes. Cluster computing is a set of connected computers with a high bandwidth network that behaves as a single unit. It can be controlled by some set of tools. (as shown in Figure 1.4). These computers are connected through a local area network and run its own instance of an operating system. Cluster computing provides high-speed computing and parallel computing in the 1980s [1]. The main advantage of these systems is that they have very cheap cost as compared with the mainframe computers and used in some small organizations. Cluster computing has emerged with a several

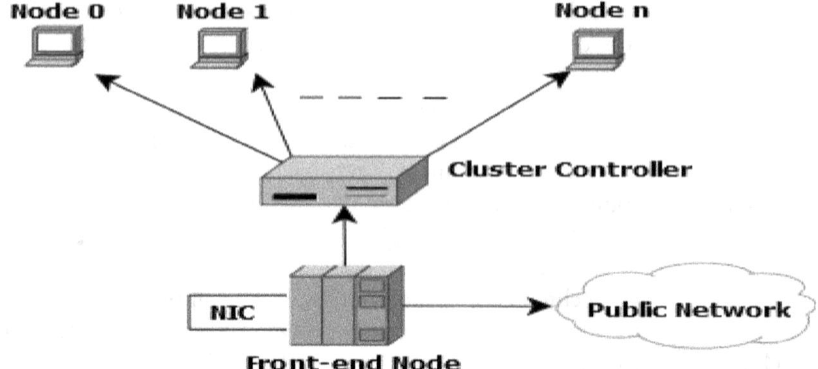

FIGURE 1.4: Cluster computing.

high-speed processors, low-cost platforms with connectivity, and software tools for distributed computing. These kinds of systems provide higher reliability due to easy detection and replacement of a faulty node. Load-balancing and high-availability clusters improve the overall performance. The evolution of cluster computing has contributed a lot toward the development of parallel virtual machine (PVM) and message passing interface (MPI). It also provides the ability to provide scalability if required with the increasing demand of computational power.

1.2.3 Grid computing

In the early 1990s, grid computing [18] was introduced as an evolution of cluster computing. It also makes use of group of physically connected computers to execute a single dedicated operation and to solve some complex problems. A grid is connected using a grid middleware software which is used to translate the information from one format to another recognizable format. Let us differentiate between the cluster and grid computing. Grid computing nodes can be dispersed geographically and heterogeneous; whereas in cluster computing, all nodes must be managed in a single location with a network. Grid computing is also known as the predecessor of cloud computing due to its processing power and distributed nature. The architecture of grid computing can be used for redundant network connection and load-balancing environment. The major benefit of grid computing is that it provides parallel processing which enables a developer to divide a program into small segments and solve the same each segment problem independently and then combine the results of each segment to produce a single solution. Grid computing can be used for a large organization where several machines are sitting idle at a particular moment.

1.2.4 Distributed and parallel computing

In the early 21st century, there were requirements of multiprocessor design and faster programs to solve complex problems. Distributed and parallel computing [19] address the problem and act as a foundational model of cloud computing that supports connectivity of multiple nodes with a network. The parallel and distributed computing provides several basic services like consistency in memory updation, concurrency, mutual exclusion, message-passing, and shared-memory. The major difference between these two is that distributed computing supports multiple processors that may be distributed in different nodes and connected via memory channel using a common network. Whereas parallel computing supports multiple processors deployed in same node and communicate with shared memory bus. There are two kinds of processor architecture in parallel computing: tightly coupled and loosely coupled architectures. Tightly coupled multiprocessors are able to share memory and communicate by information exchange among processors. In loosely coupled multiprocessors, communication can be done by sending messages to each other across the physical links. The most efficient topology of parallel computing is a hypercube, in which each node is connected directly with some neighbors.

1.2.5 Virtualization

Virtualization [20] is introduced around 40 years ago but it has limited number of applications. It has not been utilized efficiently due to resource constraints. However, these limitations have overcome and it has become the foundational element of cloud. Virtualization allows the end-user to access several computation technologies and storing components on-demand with a pay-per-use basis. The main utilization of virtualization is that it simulates the interface between hardware and end-user. Virtualization can be integrated with several latest technologies, which helps in developing a powerful computing environment. Hardware virtualization integrated with the software stack provides the platform called virtual machine instances. Several virtual machine instances can be executed on high performance computers. To replicate the runtime environment of programs, virtualization can be used. There are several types of virtualization types such as hardware, software, storage, and operating system virtualization. Virtualization is supported by specialized software such as Hypervisor which provides connectivity between the server and the virtual environment.

1.2.6 Web 2.0

A web interface is required through which cloud computing can deliver its services. Currently, the web has been evolved with several services and functionalities like interactive sharing, collaboration, the composition of application, and user-centered design. Web 2.0 [21] is the current state of online technology that provides several new features as compared to the web. It provides the better user interaction, improved channels, and collaboration. Web 2.0 is an extended and dynamic version of the web which is able to share information online through social media, Internet, and web-based communities. There are several advantages of web 2.0 such as rich web application, latest technical specification, user-friendly, and dynamic learning communities. There are several applications of web 2.0 such as Google Maps, Flicker, Facebook, Blogger, and YouTube. Flicker provides some advanced services to store digital images, videos, and online diaries. It brings interactivity and flexibility of the web pages to improve the user experience using web-based access for desktop applications. Finally, the main aim of web 2.0 is to leverage the utility of the Internet to everyone.

1.2.7 Service-oriented computing (SOC)

SOC [22] is the foundational model for cloud computing and provides support to the low-cost, flexible, and interoperable system. A good SOC model must have the following properties like loosely coupled, platform-independent, and transparent to the location. It is difficult to endeavor to adopt the SOC due to the early stage of the technology and incompleteness of web services. There are several standards for web services such as Web Services Description Language (WSDL), the Simple Object Access Protocol (SOAP), and the Business Process Execution Language (BPEL). SOC introduced two main concepts such as Software-as-a-service (SaaS) and quality of service. The term SaaS has been inherited from the application service provider (ASP) that delivers the software services across the world on-demand and rental basis. Quality of service requirements can be shared among the client and the providers. SOC relies on service-oriented architecture (SOA) to build a service model and a way to recognize applications and infrastructure. SOA is capable of the service discovery, integration, and overcome different challenges of distributed computing.

1.2.8 Utility computing

It provides computational services through an on-demand and pay-per-use basis. Utility computing [23] is a popular IT service that provides flexibility to the end-user. Users can access these services economically. Utility computing is a model that is very similar to basic service models such as electricity, telephone, and gas. The end-user can access the services virtually using a virtual private network through the Internet. The back-end services and infrastructure is managed by the service providers. There are several applications of utility computing such as grid computing, cloud computing, and managed IT services. It also includes some basic storage functionalities like virtual storage, virtual server backup, virtual software platform, online backup, and most IT

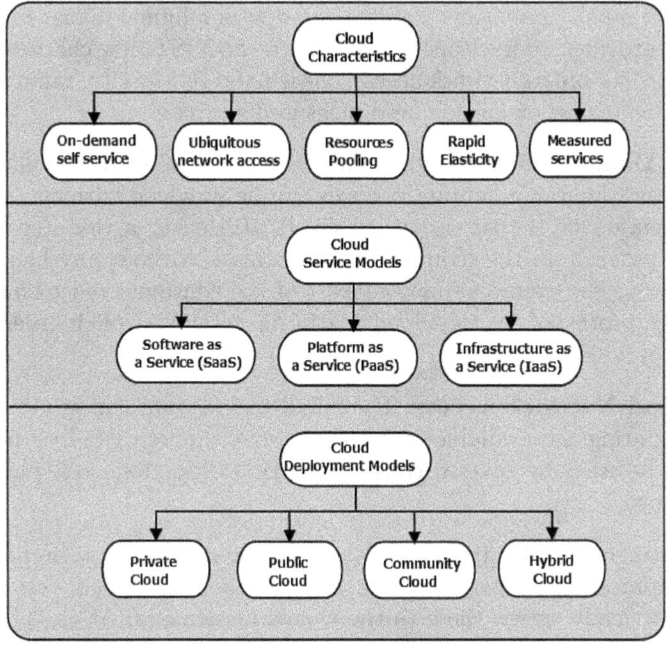

FIGURE 1.5: Cloud characteristics, service models, and deployment models.

solutions. Scalability, elasticity, virtualization, automation, and standard utility computing are some of the characteristics of utility computing. The main advantage of this model is that it reduces the cost of IT services, hardware, and optimum utilization of existing services.

Let us understand the cloud characteristics, deployment models, and service models which are explained in below sections and also shown in Figure 1.5.

1.3 Definitions and Characteristics

The US Department of Commerce introduced an agency that is responsible for providing standards in the field of Science and Technology, named the National Institute of Standards and Technology (NIST). The NIST worked continuously and after the 15th version. The NIST definition [24] of cloud computing presents, "Cloud computing is a model for enabling ubiquitous, convenient, on-demand network access to a shared pool of configurable computing resources (e.g., networks, servers, storage, applications, and services) that can be rapidly provisioned and released with minimal management effort or service provider interaction." According to the NIST, few characteristics of cloud are as [4]: broad network access, on-demand self-service, rapid elasticity, resource pooling, or expansion, and measured service.

- **On Demand Self Service:** Email, services of server, application, or network kind of computer services can be provided without explicit intervention of the service provider. It also ensures that the customer can perform all the required actions himself without any help from IT experts. For example – any request of the consumer must be automatically processed by the cloud platform, without explicit interaction of the provider.
- **Broad Network Access:** The NIST ensures that the services of cloud computing are available over the network through the Internet, which can be used by diverse consumers like IoT devices, cell phones, and laptops.
- **Resource Pooling:** To serve the multiple consumers with different geographical locations, providers used to pool the virtual resources and dynamically assign them to the clients on-demand. At the higher level of abstraction, customers must be able to specify locations like country, data-center, etc.
- **Rapid Elasticity:** To scale on-demand, the capabilities available can be elastically provisioned and released. The rapid elasticity can be disbursed to provide the quantity.
- **Measured Services:** Cloud service can automatically control and optimize resources at the level of abstraction according to the service type. The resources should be monitored, controlled, and reported to have transparency for all customers as well as provider.

1.4 Cloud Service Models

The cloud computing model is capable of providing convenient, on-demand, and ubiquitous network access to storage, and services that can be provisioned with minimal intervention. The cloud service providers (CSPs) ensure that the cloud services are having the computational capabilities which can support the essential characteristics of cloud computing. There is three kinds of possible service models provided by cloud computing such as Software-as-a-Service (SaaS), Platform-as-a-Service (PaaS), and Infrastructure-as-a-Service (IaaS) [25]. The major factors to categorized cloud services are the cloud service customers (CSCs) like end-user, IT operations, or developers, and computing capabilities like software, infrastructure, or platform. As per the NIST cloud computing reference architecture (CCRA), the services are available at the service layer of the orchestration layer. The explanations of the services are as follows.

1.4.1 Software-as-a-service (SaaS)

SaaS [26] model enables the access to cloud-based web application with the existing infrastructure of the organizations. The SaaS applications run on the CSP's side and maintained and controlled by the vendors. The services can be accessed with a paid licensed subscription, or for free but with limited access. To access the services of the cloud, one is not supposed to install any new application, infrastructure, or software. The major benefits of SaaS are affordability, ready-to-use, affordable, and accessible anywhere. There is no requirement of having on-premise hardware for the cloud service model which is opted by small-scale organizations. These services can be accessed anywhere with Internet connectivity. The applications integrated with the SaaS are ready to use services and require no set-up or installation. Although there are several benefits of this service model, some demerits are also there in the SaaS model like lack of control (full access of the vendors) and slower speed (applications are accessed through the Internet and run in remote servers). The SaaS-based cloud services have fewer features and functionalities as compared to the client-server co-relations. Although these disadvantages can be avoid if the organization deals with the offered features only.

1.4.2 Platform-as-a-service (PaaS)

A third-party vendor provides a platform through which we can develop, test, or run the applications. PaaS [27] service model has the capability to eliminate the need for in-house premises hardware and software installations. Unlike SaaS, we can control and manage the underlying cloud platform through the deployed applications. Default services are offered in terms of the servers, networking, or storage manageable by the vendors or the platform providers. The main advantages of the Paas service model are cost-effectiveness, multiple programming language support, scalability, minimum development time, and enhanced collaboration. In the PaaS model, it is difficult to switch from one PaaS vendor to another vendor, which is the major drawback of this model. PaaS model is having some security and compatibility issues with the cloud model. So, due to compatibility issues, we may be required to leave some elements out of the cloud model.

1.4.3 Infrastructure-as-a-service (IaaS)

IaaS [28] is the most flexible cloud model which provides the complete and scalable control over the customization and management with existing infrastructure. This model has the capability to replace the traditional infrastructure components and on-premise data-centers like servers, networking devices, and storage units. It provides a virtual provision of computing services over the cloud through the Internet. Amazon Web Services, Microsoft Azure, and Google engines are some of the examples of the CSPs. Minimized capital cost, simple deployment, and flexibility are the main advantages of the IaaS model. In the IaaS model, it is very easy to deploy the servers, storage, and networking for smooth functioning. However, the cost of the IaaS model is more than the PaaS or SaaS model. It's difficult to get the detailed insight details of the IaaS model due to restrictions imposed by cloud admin.

1.5 Cloud Deployment Models

Cloud services are the future of IT solutions which is all about outsourcing the services and infrastructure to access it through the Internet. The growth of cloud computing enhances the service deployment models [29] and its strategies. So, there are several enterprises of cloud solutions which depend on the degree of required outsourcing. To enable the fast-loading, there are thousands of servers and storage devices of cloud hubs. Cloud hubs are given the priorities to the user according to the closeness of the user's geographical location. Therefore, the deployment models of cloud computing can be categorized based on geographical location. A more specific categorization is provided according to the administrative domains of the cloud. The categories of various deployment models are as follows.

1.5.1 Private cloud

Private cloud [30] can be used by stand-alone organizations. It also provides control over security and backed up by a firewall that can be hosted externally or internally. These kind of clouds are implemented within the campus or a building for private use and generally can be accessible to the organization itself. Private clouds are the best solution for the organizations which require high-security and availability. It also has the possibility of testing applications and systems with low cost as compared to other cloud services. Some key advantages of private clouds are as follows:

- **Information Protection:** The first and major key advantage of private cloud is to keep the information safe within the premises of the organization. Although several public cloud vendors claim for the good quality services with high-security. There are lots of vulnerabilities which are founded by several hackers in the past.

- **Assure Service Level Agreements:** Data replication, monitoring and maintenance, recovery, and appropriate recovery mechanisms are some of the requirements which are expected from the private cloud to ensure the quality of services. Although public cloud vendors provide some of the features. All features can be achieved through the private cloud.
- **Compliance Using Procedure:** To deploy and execute some applications through any third-party services, the standard procedure and operations need to be ensured. It may not be possible for virtual public infrastructure.

These kinds of services and benefits make the private cloud more secure and available for better services. If we consider the architectural view, private clouds are heterogeneous in nature due to deployment using existing infrastructure. It could be a desktop grid, a cluster, a data-center, or a combination of them. Although private cloud has lots of features, they have limited capability to scale elastically on-demand when compared with the public clouds.

1.5.2 Public cloud

Public cloud [31] services are provided for public use through a network for general use. End-users do not have any control over the location of infrastructure. The canonical viewpoint of the public clouds says that services are available with anyone, from anywhere using the Internet. Public clouds are available as a shared cost model or pay-per-use per user basis. These kinds of model are best suited for the growing or variable requirement organizations. They offer the solution to minimize the IT infrastructure cost and can handle the existing infrastructure. There are some key advantages of public clouds.

- **Multi-tenancy:** A public cloud has a multitude of users, not as a single user and a user can work in a virtual environment separately than the other users. This is one of the primary requirements to provide effective monitoring of the user and provide quality of services.
- **Ubiquitous Model:** A public cloud is able to offer all three service models such as platform (for example, Google AppEngine), infrastructure (for example, Amazon EC2), and Software (for example, salesforce.com).
- **Less Restrictions:** If we consider an architectural view of public clouds, there are very few restrictions. Public clouds can be composed of geographically dispersed data-centers to share the load of users and to serve server them according to their locations.

These kinds of clouds are having several benefits but less secure due to availability for the general users. Public clouds also provide a wide range of services to the end-user with the existing infrastructure. So, there is no need to set-up high configuration hardware or software to start a new organization.

1.5.3 Community cloud

Community clouds [32] are distributed systems that can be created with the help of distinguishing cloud integration to address the specific requirements. Basically, it is a mutually shared model among several organizations which are belonging to a particular community like government agencies, banks, scientific research, and commercial organizations, etc. These models of cloud can be developed by an internal employees or third-party vendors by multiple administrative domains. Some key benefits of community clouds are discussed as below:

- **Control and Convenience:** Community clouds have enough flexibility to design it as per the requirements. So, there is no conflict between control and convenience that can make all decisions through a collective effort.
- **Environmental Sustainability:** Community clouds are supposed to emit less carbon due to the under-utilization of resources. So, these clouds tend to provide more environmental sustainability.
- **No Single Point of Failure:** There are multiple administrative hosts that provide infrastructure and support to the community cloud. If there is one system failure, then the remaining system will work properly.

A community cloud can also be formed as an aggregation of the resources of several communities. Community clouds are heterogeneous in nature and independent from the vendor's agreements. These clouds are also known as open systems due to that fair competition among different solutions.

1.5.4 Hybrid cloud

Hybrid cloud [32] consists of a large hardware and software infrastructure that can fulfill the requirements of multiple users. This model combines the best practices of a public and private cloud but as separate entities. A hybrid cloud can provide scalability, security, and flexibility to the end-user. There is an ideal situation of the organization in which an organization uses a private cloud within the organization and uses the public cloud to interact with its customers. There are some key points about hybrid clouds:

- **Scalability:** Hybrid clouds can address the scalability issues and can provide distinguished clouds for different services. For example, to deal within the premises and outside of the premises.
- **Cloud-Bursting:** In the hybrid cloud, we can have several services or resources for a time period till required which can be released after usage. This kind of flexibility is provided by the hybrid cloud and is known as cloud-bursting.

- **Infrastructure Management:** Hybrid clouds provide the facility of infrastructure management software like OpenNebula which is a kind of scheduler that provides cost-based scheduling.

The provisioning of services can leverage the distinguished other services to ensure the quality of services. There are several other benefits of hybrid cloud such as higher efficiency, faster deployment, low upfront cost, and simple to manage infrastructure.

1.6 Cloud Service Platforms

In this section, various cloud service platforms are explained which are offered by various cloud vendors.

1.6.1 Amazon web service (AWS)

Amazon Web Services began its IT infrastructure services for various firms which are provisioning web services. These services are known as cloud computing nowadays. The main advantage of this kind of services is that they replace capital expenditure to operational expenditure (low-cost paradigm). AWS provides a scalable, reliable, and low-cost infrastructure in the cloud that provides hundreds of businesses around the world. AWS provides computation services and various other services to help grow an organization. The services are based on several protocols such as HTTP, REST, and SOAP protocols.

1.6.2 Microsoft azure

Microsoft Azure [33] service is provided for Windows platform that mainly have three components that provide particular services to the end-users such as windows Azure, SQL, dot NET, and azure services. The services run on the cloud servers. For example, the SQL server is offered as a service by SQL azure. The local applications are running through the .net services. There are distinguish services which are offered for the cloud users like virtual machines, identity, storage, mobile services, data management, and messaging, etc.

1.6.3 Google cloud platform

Google provides a cloud platform which is named as Google Compute Engine (GCE) [34]. GCE is the IaaS component that is used for global infrastructure which runs different services such as Gmail, YouTube, Google search engine, and other services. A GCE instance can start with a disk resource that is known as persistent disk. GCE provides the facility to the end users to launch the virtual machine on demand that can be customized according to their requirements. It provides several features like VM performance, transparent maintenance, billing, pricing model, and global scope for images and snapshots. Google VMs can boot within 30 seconds that is about 4–10 times faster than other VMs.

1.6.4 IBM cloud

IBM provides cloud services [35] which is a public platform with different products like storage, computation, networking, development, testing, security, etc. The operation and management of these services is done by IBM. The IBM cloud platform combines IaaS and PaaS and integrates the services of infrastructure with different platforms. This cloud platform supports both large and small organizations. IBM cloud services are built to support the requirements of public cloud or multi-cloud model. It provides different open-source technologies such as Kubernetes, computation options, Red Hat Openshift, VMs, or containers. Cloud native applications can be deployed to ensure workload portability.

1.6.5 Adobe creative cloud

The video editing, designing of the graphics, web development, and many more services are the collection of software services which are provided by the adobe creative cloud. Adobe creative cloud [36] is available on monthly or annual subscription and delivered via Internet. It retains several features of adobe creative suits with new features like instant upgradation, easy-sharing, storage to the cloud, etc. There are some services provided by the adobe creative cloud such as spark, premier rush, XD, Fonts, and portfolio. It is a tool which enables both collaboration and creativity. It builds fully functional websites from ground data to mobile designing. Creative cloud supports both MAC OS and windows version of the system.

1.6.6 Kamatera

Kamatera [37] is a part of global cloud service provider which contains a rich set of services for all kind of organizations. It uses most advanced technologies with high level of customer services. Kamatera is operating thirteen global data-centers and serving thousands of customers including application developers, international enterprises, SaaS providers, etc. It provides a rich set of services such as cloud server with web hosting, wordpress server hosting, storage, cloud private network, virtual private servers, and many more. Kamatera clients can customize the services of the company according to their requirements and scale their services on hourly or monthly basis.

1.6.7 VMware

VMware [38] is a virtual-machine platform which is an abstraction of x86 PC hardware to execute the multiple operating system in an unmodified way. It indicates that multiple deployments are possible for desktop applications without rebooting or partitioning. VMware cloud comprised SaaS and IaaS which is ideal solution for application service providers (ASPs), Internet service providers (ISPs), and PaaS. The multi-cloud solutions can deliver a cloud operating model for all kind of applications. It is a world's leading public cloud which provides protection and scaling for vSphere-based applications. VMware reduces the overall operational overhead expenses and achieves faster cloud strategy in terms to leveraging the existing skills. It is still continuing to invest on people, innovation, employee productivity, and business.

1.6.8 Rackspace

Rackspace [39] offers cloud backup and block storage. Cloud block storage is released in the year 2012 and powered by OpenStack service. The Rackspace cloud provides cloud-based products and services, which offers cloud storage, virtual private server, load balancer, backup, monitoring, and databases. Rackspace cloud services deliver the innovative capabilities which increase efficiency, and generate new revenue streams. It provides distinguish services such as management of hosting, professional services, security and compliance, business intelligence, and application managed services. It is used to deliver high performance by using solid-state drives and hard drives. Whereas backup services provide file-level backups and compression techniques to improve security.

1.7 Challenges Ahead

Although cloud computing has gained lots of popularity these days due to its adaptability by the industries during a short time period. However, the research in the cloud is still in its initial stage. There are a significant amount of issues that are still needed to be addressed. Several new challenges also keep emerging for industrial growth. In the current section, we discuss some key research challenges of cloud computing.

1.7.1 Virtual machine migration

To balance the load of the data-centers in the cloud computing, virtual machine migration comes into picture. It also enables the CSP to provide the highly responsive and robust mechanism in the data-centers. This process has been evolved from the process migration techniques, which are used to implement the migration in real-time. Authors of VMware [40] have performed some live migration of VMs which can be implemented in a part of the milliseconds. Another team of authors [41] is able to migrate the entire OS and its applications as a single unit which is able to avoid several problems like migration at the process level and VM level. The main advantage of VM migration is to avoid the hotspot in the real-time, which helps to detect workload hotspot. The secure VM migration is an evolving research area in which research is still going on.

1.7.2 Interoperability and standards

Cloud computing is a kind of service-based model to provide infrastructure such as water and electricity. For the proper utilization of the services, several vendors want to provide interoperability and standards between distinguish solutions. But, vendor lock-in is one of the major obstacles against the growth of cloud services. If an organization wants to switch its CSP, then it requires a considerable amount of time, conversion cost, and resources. Hence, the presence of interoperability and standards provide a room to choose the vendor and switch them easily. There are some organizations that are leading the path to standardize the services of cloud computing such as Cloud Computing Interoperability Forum (CCIF), DMTF Cloud Standards Incubator, and Open Cloud Consortium [42]. Another approach to provide the standards to the cloud is a general reference architecture and a standard interface through which a user can interact.

1.7.3 Security and privacy

A secure communication among cloud nodes [43] helps in meeting the confidentiality and privacy of user's data. To achieve the high-end security among the data-centers, good security measures should be in place. One cannot rely on CSP for providing better security solutions. The providers must achieve the following objectives such as auditability as a check point and confidentiality to secure access. Confidentiality factor can be implemented through the use of encryption techniques, whereas the auditability can be ensured by using remote attestation techniques. These techniques need a trusted platform module as a proof of security. However, remote attestation techniques are not sufficient. Hence, there is a need to build a trust mechanism at each layer of the cloud. The development and deployment of efficient and robust Intrusion Detection System (IDS) is another security challenge in cloud. Specially, some of the intrusions may target hypervisor layer to compromise the entire system. Introspection-based approaches play a vital role to provide cloud-specific security solutions. However, not much research work has been done in this area.

1.7.4 Energy management

There is another issue with the cloud services called as management of energy. The 53% of the total expenditures of the data-centers have invested in powering and cooling of the system [44]. Hence, CSPs have more pressure to reduce this cost and improve the energy efficiency for the operational centers and cut down the total cost. It attracts the researchers to improve the energy-efficient techniques and meet the requirement of government standards as well. There are several solutions to reduce the power consumption provided by several researchers [45] like the architecture based on the energy, scheduling of jobs for energy-aware, and the protocols used for the networking purpose. Although several techniques have been proposed by authors. However, it is more challenging to achieve a trade-off among the application performance and energy-efficient mechanisms.

1.7.5 Accessibility issues

The access control applies to the read and write allowances for [46] authenticated users. The username and password are used to provide the authentication of the system. In the multi-tenant cloud environment, there are a large number of customers. A significant number of customers are in the multi-tenant cloud world. Each client uses website or front-end GUI to access cloud services. Therefore, distinguished and efficient access control techniques are required to be developed to solve the authorization issues.

1.8 Conclusion

Cloud computing has gained popularity in recent time to manage and deliver services through the Internet. Cloud computing can provision applications, storage space, and several software services as per the demand of users. The ultimate aim of cloud computing is to deliver the services as pay-per-go manner just like basic services such as water and electricity. In fact, a small industry or start-ups can initiate their work without any pre-defined hardware or software requirements. However, despite of having significant advantages provided by cloud computing, there are several key challenges which are still not covered by researchers like energy management, security, trust, interoperability, etc. We also provided some few definitions with the discussion on history of cloud computing. Various key technologies have also been surveyed with the emergence of cloud. The standard definition with cloud characteristics has also been covered in the current chapter. Various cloud service models, such as IaaS, PaaS, or SaaS have also been discussed along with cloud deployment models such as private cloud, public cloud, community cloud, and hybrid cloud. At the end, several research challenges are discussed which provides future research directions.

1.9 Questions

Fill in the blanks

1. Arrange the development of following in ascending order:

   i Microsoft Azure
   ii Hadoop
   iii Google cloud
   iv Amazon web services

Mark the correct option for answering the question 1.
(a) iv, iii, ii, i (b) iii, iv, ii, i
(c) ii, iii, iv, i (d) i, ii, iii, iv

2. Which of the following is incorrect statement:

   i Applications of web 2.0 are Google map, Flicker and Facebook.
   ii A service-oriented computing model must be tightly coupled and platform dependent.
   iii Virtualization can be integrated with several latest technologies.
   iv Utility computing is used in grid computing and cloud computing.

Mark the correct option for answering question 2.
(a) i (b) iii
(c) ii (d) iv

3. Cloud Computing is used for

   i Infrastructure provisioning
   ii Platform provisioning
   iii Database provisioning

Mark the correct option for answering question 3.
(a) i is true (b) i & ii are true
(c) i, ii & iii (d) none of the above

4. A cloud environment consists of

   i Cloud Controller Server
   ii Cloud Compute Server

iii Cloud Network Server
iv All above mentioned

5. Cloud bursting is common in

i Hybrid Cloud
ii Private Cloud
iii Public Cloud
iv All above mentioned

Short-Answer Questions

1. Define cloud computing and describe various characteristics of cloud computing.
2. Explain various service delivery models of cloud computing with examples.
3. Discuss various cloud computing deployment models with example scenarios where they are most suitable for.

Long-Answer Questions

1. What is the need of cloud computing. Discuss all the architectural components of cloud architecture with suitable diagram in detail.
2. Discuss how cloud computing is different than traditional computing environment. Explain various open research challenges in cloud.

Chapter 2

Introduction to Cloud Security

2.1 Introduction

Cloud security consists of set of technologies, controls and policies designed to protect the applications, infrastructure and data of cloud environment. It can also be considered a sub-branch of computer security and network security. It consists of the security constraints designed to incorporate the cloud service provider and end-user perspectives. The importance of cloud security has been immensely increased in the modern computing era. There are several users who are gradually adopting cloud for hosting their applications and data. However, there are still various concerns which prohibit the users/enterprises/organizations to adopt the cloud-based infrastructure. Cloud Security Alliance (CSA) did a survey in the year 2019 and found that security is the major concern for majority of users [47].

CSA's report stated that a total of 81% users have the security concern while adopting the public cloud platforms. Whereas 62% users are worried about data loss and leakage risks and 57% people are worried about the regulatory compliance as shown in Figure 2.1 [47]. They found 49% users have concern for issues related to integration with rest of non-cloud IT environment while adopting the public cloud infrastructure. Around 44% users have legal and cost related concerns. A total of 39% users are worried about the visibility issues and 35% emphasize more on application-migration related cloud adoption concern. There were 32% users who have concern related to lack of expertise staff to handle the cloud services. There also stated that 23% users have concern of not having a staff to manage cloud services. Only 2% cases were reported for having vendor lock-in related issues.

The flexibility and easiness of cloud services have opened doors for attackers. The attacking incidents which are happening in the cloud-based IT environment, raises a big question for securing cloud environment. Some of the security agencies have reported various attacks such as Virtual Machine Escape, discovered by research outfit VUPEN security [48] in 2012. This attack affected the error handling function of Intel processors. According to the report of European Network and Information Security Agency (ENISA) [49], Dropbox has been affected by Distributed Denial of Service (DDoS) attack. The DDoS botnet was also launched against amazon web services. Hackers

DOI: 10.1201/9781003004486-2

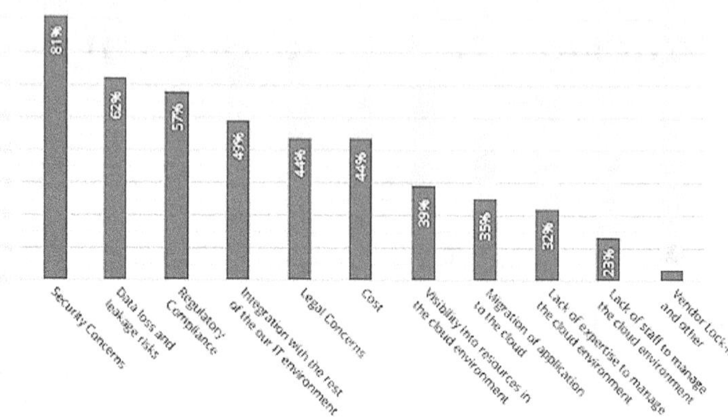

FIGURE 2.1: Concerns shown by enterprizes/user while adopting public cloud infrastructure.

used the exploit in ElasticSearch and attacked the amazon EC2 instances in 2014 [50]. It has also been reported by cyber threat defense that 75% attacks apply the known vulnerabilities which are still present in software in 2014 [51]. Infact Code Space is attacked by attackers, causing destruction of the customer's sensitive data in 2014 [52]. Symantec reported that 494 vulnerabilities and two zero-day vulnerability in 2015 [53].

Internet Security Threat Report stated the proportion of evasive-malware samples in cloud environment that can detect the virtualization environment as shown in Figure 2.2 [54]. Cisco does the recent survey and classified Trojan as top five malware attacks to gain access to user's computer and organization network [55] in 2017. It is one of such attack which gains access to user's computers and network. The investigation carried out by cisco reveals that 75% organizations are victim of malicious software which can be used to launch advanced attacks. As per CSA report in 2018, there were a total of 23420 phishing website links were found which have very high compared to 2016 report [56]. Recently, a cloud service provider company "iNSYNQ" became a victim of ransomeware attack which caused shut down of their network system, making it difficult for customer to access the services on 2019 [57].

The increasing attacking incidence each year raises a concern for security in cloud environment. Below, some of the key vulnerabilities in cloud are discussed which are usually exploited by attackers as shown in Table 2.1.

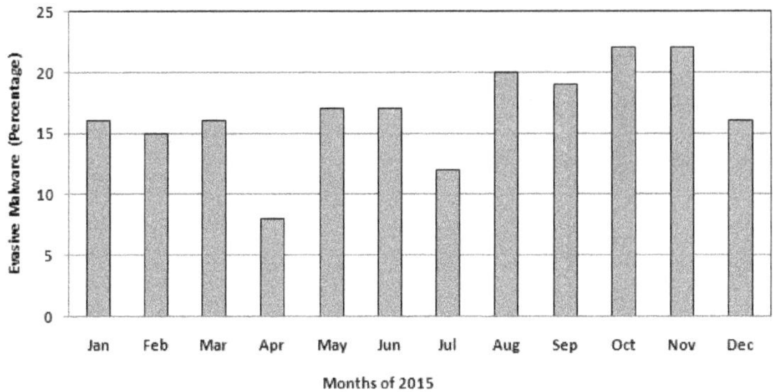

FIGURE 2.2: The attack statistics of virtualization-aware evasive malware samples in cloud environment.

TABLE 2.1: Key vulnerabilities in cloud environment

Vulnerabilities	Consequences
Lack of physical control	Loss of physical control on the data located in remote cloud servers can lead to data leakage and modifications, etc.
Under-provisioning of bandwidth	The under-provisioning of bandwidth can lead to consumption of resources by causing flooding attacks such as DDoS.
Pricing model of cloud	Compromising the billing model of cloud can incur in generating the incorrect billing information, causing disputes between CSP and tenants.
Insecure Browser and APIs	Breaching the securing control through insecure APIs can cause unauthorized access to resources
Illegitimate access to management interface	Attackers can gain the control on the management console and gain access to administrative access.

2.1.1 Vulnerabilities present in cloud

Let us now understand various major vulnerabilities in Cloud Computing.

Vul1 – VM Co-tenancy

The co-tenancy, also called as co-residence refers to sharing of same physical resources by different cloud customers or tenants. Cloud computing provides the better utilization of resources through this concept in which different tenants may have their VMs, running in the same physical machine. It raises a serious concern about security. Such a vulnerability can be exploited in launching attacks such as VM Escape attack [48], Cross-VM side channel attack [58], etc.

Vul2 – Lack of Physical Control

In cloud computing, the tenant's data, program and their computation is outsourced to remote servers of cloud. This leads to lack of physical control on tenant's data, programs and machine, etc., which may result in dangerous attacks [59]. For example, the tenant's data residing in remote cloud servers can be modified, leaked or even lost. An attacker can inject the malicious code in the tenant's software application running in cloud and cause harm in cloud resources. It could lead to dispute between cloud service provider and tenants. It is difficult to ensure the data confidentiality and integrity just by making use of traditional security mechanisms.

Vul3 – Under-provisioning of Bandwidth

Traditional DDoS attacks have been prevalent in cloud computing environment as well in which network layer is flooded with excessive traffic connections to exhaust the resources. In addition, a new form of DDoS, exploits the feature of under-provisioning of resources in cloud environment. In this form of DoS, shared resourced such as memory, storage and computation is consumed by attacker excessively, making the other VM users deprived from resources. DDoS in cloud, also takes the benefit of under-provisioning of cloud resources. CISCO's states that a data center design is under provisioned with a factor of 2.5:1 to 8:1 which means the network capacity of data center is less than the average capacity of hosts running in same subnet [60]. The under-provisioning of bandwidth exploited by attackers in flooding the network.

Vul4 – Pricing Model of Cloud

Cloud Computing is based on the pay-as-you-go pricing model. The pricing model is based on the computation of metrics such as bandwidth consumed, servers/VM hours, storage capacity, bandwidth, CPU utilization, etc. However, if pricing model is compromised, the billing information generated by cloud service provider will provide the incorrect information [61]. Tenants will suffer by paying for additional billing charges. One such attack is called Economic Denial of Sustainability (EDoS) attack [61] which affects the pricing model of cloud adversely.

Vul5 – Insecure Browser and APIs

Cloud vendors provide a rich software API support in order to allow customers to interact and manage with services. The APIs helps to effectively perform the service provisioning, service orchestration, service usage and service deallocation/release through web browser (client program). Any vulnerability present in the APIs can be used by attackers in breaching the security of cloud services. There are many web-based attack possible such as phishing attack, SSL certificate spoofing attack, cross-site scripting attack (XSS) and sql-injection attack, etc. [62]. Web Security standards should be followed by the APIs in order to ensure the safety of services though web-based attacks.

Vul6 – Illegitimate Access to Management interface

A management interface such as AWS management console [63], is used to manage the client's subscription for resources such as instance, storage, computation, etc. If the management console is breached by unauthorized users, it may lead to drastic consequences. Cloud environment consists of more number of users and administrators when compared to traditional computing environment. It raises the probability of gaining unauthorized access to the cloud resources through a weak management interface. Insecure cryptographic keys and algorithms can also result in such security breach. For example, Amazon EC2 management interface was breached in the year 2011 through XSS attack which exploited a cryptographic loop hole in service.

Vul7 – Insecure Internet Protocols

Any algorithmic vulnerability present in internet protocol can also be helpful for attackers in bypassing the cloud security mechanism. ARP poisoning is one such example of protocol vulnerabilities which can be exploited by a malicious VM user. A malicious user can exploit the vulnerability in re-routing the traffic of some co-located VM to malicious VM. Some other attacks which exploit protocol vulnerabilities are RIP attack, DNS poisoning attack, flooding attack, etc. HTTP protocol is also vulnerable to session-hijacking session-riding and used as web application protocol for web service running in cloud [64].

2.1.2 Need of cloud security

Cloud computing has become a successful business model because of the ease and flexibility in provisioning and managing services. However, the lack of control on data and services has raised serious concerns related to security. Therefore, there are still many enterprises who fear in migrating their services to cloud. In traditional data centers, there was a direct control on every layer in the computing stack from hardware to software. However, in virtualized data centers, the direct control is lost and complete stack comes directly under the control of cloud service provider. Hence, an organization is required to have a significant level of trust with cloud service provider before transferring services into cloud. Security has been the major barrier in adoption of cloud services over the years [47].

There are three different types of cloud service models in cloud, i.e. SaaS, PaaS, and IaaS. The vulnerabilities and security requirements for each layer varies from layer-to-layer. SaaS provisions the web-based services in most of the cases through internet. Attackers target the API interface of the web portal and sometimes the secure shell (SSH)to launch attacks in SaaS services by hijacking private keys, credentials and API keys of the tenants. The web interface can be exposed to cross site scripting attack (XSS) and signature wrapping attack [62].

Zhang et al. [65] proposed an attacking framework to launch cache-based side channel attack in PaaS cloud. The victim's VMs execution is traced by providing the extended version of Flush-Reload attack. The tenants co-location is traced first and then the private key information is extracted by the proposed attacking framework. Other form of PaaS attack targets the VM images and injecting malicious code resulting into data stealing [66]. IaaS infrastructure can also be subjected to various attacks.

The cloud infrastructure can be subjected to various malware and network attacks. For example, the virtual switch which connects the co-located VMs, can be sniffed to monitor the network traffic of co-located VMs. The virtual network can also be subjected to ARP spoofing, MAC spoofing attacks. ARP spoofing is used to reroute the traffic to malicious VM by alerting the ARP table. MAC spoofing is used to imitate another host in the same network [67]. The pricing model of cloud can also be compromised, leading to incorrect billing report generation for the infrastructure usage or software usage or platform usage.

CSP usually provides maximum access privileges of VMs to the cloud customers/tenants. A malicious tenant can attempt to gain root access of machine by installing malicious software such as rootkits. Once root access is obtained, attacker can try to breach the security of hypervisor or even the host operating system. In addition, nowadays, attackers are making the malware code more advanced which senses the monitoring virtualized environment and even the security analysis tool. On detection of the analysis environment, malware changes its behavior and bypasses the detection approach. Hence, traditional security techniques may fail to detect such advanced malware. A compromised VM can be used to launch further attacks in cloud and is a big threat to cloud. It may ultimately lead to monetary disputes between service provider and service consumers.

One of the key features of cloud is portability and inter-operability of services. However, the live migration of services can be exposed to serious security and privacy threats. Man-in-the-Middle (MITM) is one such threat in which intruder can eavesdrop the connection and mimic the destination machine [68]. The sender machine will then communicate with attacker machine, considering it as destination machine. Some of researchers, who are working in the area of cloud security, have proposed security frameworks [69, 70] to deal with security threats in cloud. However, a centralized security framework becomes a bottleneck when there is an increase in the number of TVMs in the cloud host and when the security tool uses centralized resources.

A distributed security framework that deploys the same security solution at all layers in cloud becomes less efficient because of the limitations associated with different layers. One of the security solutions could be monitoring specific security-layer(s) of cloud which will be centrally controlled and configured by Cloud Service Provider (CSP). It will enable the CSP to assign the specific security solution based on the tenants demands. In some of the existing frameworks, the security solution is deployed in individual tenant virtual machine which will be controlled by the administrator for security reasons. The chances of subversion of security tool, in such a scenario is quite high. Moreover, the security becomes much costly as individual security daemons are developed and deployed in each of the virtual machine. Moreover, the virtual machine-level solutions cannot be directly applied to the hypervisor-layer. Research are working toward introspection approaches for providing the security from hypervisor-layer.

The shared and distributed nature of cloud is complex and prone to various security challenges. The implementation of the security solution by cloud customers is not applicable to the management level behind the virtual infrastructure. Although a service level agreement (SLA) is signed between service provider and service consumer and major security and privacy related aspects highlighted. However, there is still no standard design methodology for SLA [66]. Moreover, in present scenario, customers are still least aware about the attacking incidents and open vulnerabilities and reports. There is a strong need to work toward cloud security area to provide more efficient, transparent and distributed security solutions to deal with various cloud threats.

2.2 Cloud Security Concepts

It is important to understand some important security concepts in cloud before understanding the security solutions. Some of the important security concepts associated with cloud are multi-tenancy, virtualization, data outsourcing, trust management and meta security, etc. The details for the same are given below.

2.2.1 Multi-tenancy

Multi-tenancy enables the tenant users to share the running instances. The sharing of single cloud platform improves the efficiency of the system. In case of IaaS cloud providers, multi-tenancy refers to the sharing of Virtual Machine Monitor (VMM) among multiple VMs. In case of PaaS cloud providers, multi-tenancy allows the users to share the same developing platform such as Java Virtual Machine (JVM) and .NET platform. In case of SaaS provider, it enables the provider to share the application software among multi-tenant users. It's easy to maintain, configure and manipulate the data stored in the single database. On one hand, multi-tenancy provides above benefits to the provider; however, it also expands the threat model and can be exploited using co-residence attacks as single server is being shared by multiple VM users. Cross VM side channel attack is one of such attack which attacks the co-located VMs by exploiting information of side channels such as cache, power, heat, etc. [71]. Denial of Service (DoS) is another example that can be launched

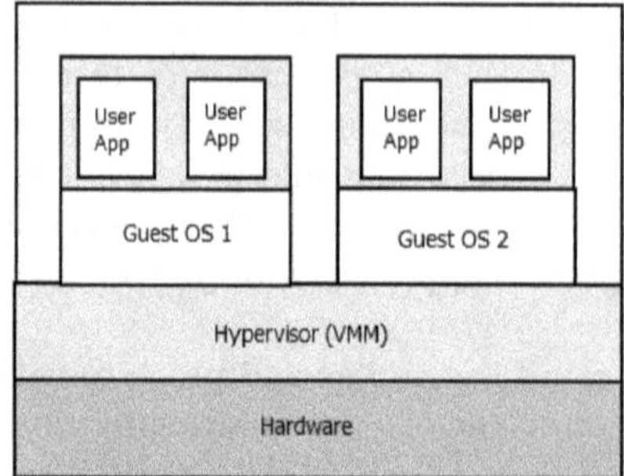

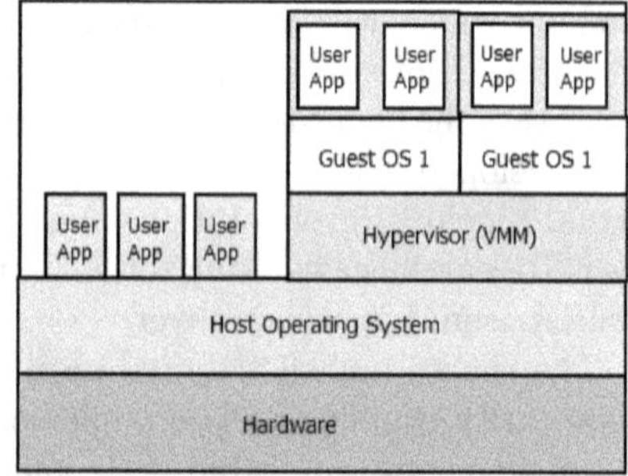

FIGURE 2.3: Types of hypervisor.

against co-located VMs very easily. VMM DoS can even consume the resources of the underline sharing platform [60].

2.2.2 Virtualization

The key technology in cloud environment is virtualization that powers the cloud environment. Virtualization enables the extraction of computing resources, services, operating system and applications from underline infrastructure on which they run. Two key components of virtualization are virtual machine (VM) and hypervisor/virtual machine monitor (VMM). A VM basically represents the emulation of the physical resources which runs an operating system called as guest OS. The emulated devices such as virtual RAM, virtual disk, virtual network interface card (vNIC) card provide the same functionalities of physical devices. A guest OS can host different applications and does not have direct access to hardware. VMM or Hypervisor runs above the hardware or software and hides the complexity of physical hardware. It allows the execution of multiple guest operating systems (OSes) in same machine. Hypervisor can easily create, delete and run different VMs having different OSes installed which is an essential requirement to provide the elastic and on-demand services in cloud computing. There are two types of hypervisors: Bare Metal Hypervisor (called Type I) and Hosted Hypervisor (called Type II), as shown in Figure 2.3. In former case, VMM can directly run on top of the hardware and access resources. Hypervisor is booted first and have access to the real device drivers. Xen [72], VMware ESX/ESXi [73] are some examples of Type I Hypervisor. In later case, it is host OS which is booted first and at the time of launch of first VM, hypervisor is loaded post-boot. The Hypervisor runs above the host operating system as a user space application. It shares device drivers from host OS to handle the input-output and completely depends on host OS for its operations. VMware Workstation [74] and Oracle Virtual Box [75] are examples of type II Hypervisor. Gradually software were developed for implementing cloud computing platform such as

Open Nebula [76], VMware vSphere [73], OpenStack [7], Citrix XenSever [72], HP Helion Eucalyptus [77], etc.

2.2.3 Data outsourcing

Data outsourcing is one of the key benefits provided by cloud service provider. It refers to the transferring of the computing, security, and especially storage to off-premise third party organization which controls the off-premise infrastructure. The capital expenditure (CapEx) along with operational expenditure (OpEx) is reduced to a significant extent. Due to outsourcing, customer loss their physical control over data which is one of the most important security issues in cloud computing. Data outsourcing also causes privacy violation as the data is outside the physical access of customers. In order to resolve this issue, customers need to be very careful while selecting a trusted service provider. Habib et al. [78] have addressed the issue of how to select a trusted service provider. A CSP should achieve the important security goals such as confidentiality, integrity and availability, etc. There are some approaches to preserve the privacy of outsourced data from outsiders. For example, Slamanig et al. [79] proposed an approach based on accumulator which maintains the Access Control Lists (ACL) having different permissions (read, write, and delete) for different users. The owner of the data can only grant/delete/modify access to the data for outsiders. In this way, unauthorized access to outsource data will be denied by the server. Another scheme called privacy-enhanced access control for outsourced data is proposed by Raykova et al. [80]. In their approach, coarse-grained access control is combined with fine-grained cryptographic access control. Adding cryptographic access control policies overcomes the trust issues with outsourced data.

2.2.4 Trust management

Trust is one of the crucial security issue concepts. It is very important for a service provider to build a trust between the tenant users and service providers. Tenant's data resides on off-premise datacenters which is completely outside the control of users. This loss of physical control raises a concern of trust on service providers. It is a multi-phased phenomenon to evaluate the trust. The security of tenant's data relies on the security management policies implemented by service provider. Tenants have to trust on them. In addition, trust on underline infrastructure of cloud such as operating system, guest OS image, hardware, application software, cloud network is another important concern for the tenant users. A trusted third-party (TTP) can authorize, audit the sensitive data of tenants and provides the security from illegitimate users. However, TTP can be compromised which leads to user's sensitive data on danger. Trustworthy systems are also discussed by Yasinsac and Irvine [81]. Tenants have to trust that systems will work fine for different circumstances such as operational error, system error, human intervention error, etc.

The trust-based systems ensure the system security as well along with the business continuity of the organization.

2.2.5 Metadata security

Cloud Organizations also maintains the massive amount of metadata which is called “data about data”. It contains sensitive information in different format. For example, Web Service Description Language (WSDL) [82] is one of the examples of metadata. An attacker can exploit the WSDL and modify it. This may cause the leakage of the user's confidential data. There are some security concepts associated with it such as data sanitization, data separation, data location and data maintenance. Data sanitization refers to permanently destroying the piece of data stored in different locations in cloud. It is a irreversible process of destruction of data. During removal, some of the metadata may not fully deleted resulting data leakage of user's sensitive information. Data separation causes the tenant to use the data of his/her domain and denial of access to other domain's data. However, data separation is done in the same hard disk being shared among multiple tenants. A poor implementation of separation policy will result in data leakage. Cloud offers data migration with VMs whenever required. The mobility feature may sometime result into loss of sensitive information and also increase the chances of errors in metadata. Man in the Middle Attack (MITM) [83] is one such threat to data in transit. It may also be dangerous to keep data at multiple-locations for backup. Data maintenance is another important security concept. Maintaining the metadata along with the applications is another challenging task. Sometimes, security attacks can take place during updation of the software patches.

2.3 Cloud Security Standards

Various unwanted attack incidents happening in cloud environment, have led to the generation of security standards. Various security standards have been discussed which covers different cloud security aspects. As per the author's assessment, various cloud security standards are as follows.

2.3.1 Information technology infrastructure library (ITIL)

ITIL [84] is security management framework that identifies best guidelines and practices which defines the process-based integrated approach for managing the cloud information technology services. ITIL is applicable to all type of IT services including the cloud services. ITIL makes sure that proper security care is take at all three levels of business operations, i.e. strategic level, tactical level and operational level. ITIL provides best practices for information security process that can be modified and used by any IT organization. A framework with continuous improvement is provided which can be aligned as per the changing need of the IT services. As cloud computing is such type of continuous changing organization. Here, the security guidelines and practices must be modified dynamically as per the business need. ITIL breaks down the information security practices into various levels:

1. Policies: The key objectives aimed by the organization to achieve. item Processes: What guidelines to follow to meet the objectives?

2. Procedures: How to distribute the activities among people and settling down the important deadlines.

3. Work instructions: What are the instructions to be performed for doing specific activities.

The key challenge faced by the organization while adapting ITIL is to redefine the set of ITIL processes that they are having efficiently. The organization is supposed to identify the gaps in the existing security processes and then mapping them as per the ITIL framework.

2.3.2 Control objectives for information and related technology (COBIT)

COBIT [85] is another security standard which is developed by international professional association ISACA which provides best practices for IT management and governance. It acts as an interface between processes and business goals. The model can also be used together with more standards such as ISO/IEC 27000 and ISO/IEC 20000.COBIT includes the following components:

1. Process descriptions: It focuses on the having a reference process model and a common language in an organization. The reference model maps the responsibility areas of planning, building, running, and monitoring.

2. Control objectives: A set of high-level requirements are provided which are to be implemented by management for having good control on IT Processes.

3. Management guidelines: The guidelines helps in measuring performance, setting up common objective, assigning responsibilities, and mapping relationship between processes.

4. Maturity models: These models are used to measure the maturity and capability for each process and identify the gaps.

2.3.3 ISO/IEC 20000

ISO/IEC 20000 [86] belongs to internationally well-known standards for IT service management not limited to cloud services. It reflects the best guidelines within ITIL. This standard is developed by ISO/IEC JTC1/SC7 in 2005. Overall, there are thirteen parts for this standard. Various requirements to establish, implement, maintain and continuously improve the service management system (SMS) are specified in ISO/IEC-20000-1 :2018. The applications of SMS as per the requirements in ISO/IEC 20000-1:2011 have been provided by ISO/IEC 20000-2: 2012. The scope definition and usability of ISO/IEC 20000-1 in applications, is being given in ISO/IEC 20000-3. ISO/IEC 20000-4 for process assessment has been withdrawn and a new series of documents have been developed by ISO/IEC JTC1/SC7. To achieve the requirements of ISO/IEC 20000-1, the best practices have been defined by ISO/IEC 20000-5. The requirements for assessment has been defined by ISO/IEC 20000-6:2017. The integration of SMS and quality management system and/or information security management system (ISM) have been defined by ISO/IEC 20000-7. The standard ISO/IEC 20000-9 has been withdrawn. ISO/IEC 20000-10:2018 defines the concepts of ISO/IEC 20000 and identifies the relationship between ISO/IEC 20000 and other standards. ISO/IEC 20000-11:2015 defines the relationship between SMS and ISO/IEC 20000-1. ISO/IEC 20000-12 defines the relationship between SMS such as CMMI and ISO/IEC 20000-1. The relationship between SMS such as COBIT and ISO/IEC 20000-1 is defined in ISO/IEC 20000-13.

2.3.4 Statement on standards for attestation engagement (SSAE)

SSAE [87] is standard developed specially for auditing tasks which is applicable to cloud service providers as well. The total 16 audits come under one of the three classes of Service Organization Controls (SOC): SOC 1, SOC 2 and SOC 3. SOC 1 focuses on financial reporting control. SOC2 assesse the security control of technical and operational tasks and defines the trust services principals for them. SOC 3 ensures and reports whether the service provider is meeting the principals of trust services. The reports generated by SOC3 are freely distributed. NIST provided a Cybersecurity Framework (CSF) which focuses on applying the federal security assessment and security authorization controls in industries owning their own infrastructure. CSF is a standard security framework for private cloud service providers.

2.3.5 Cloud security alliance (CSA) cloud controls matrix

CSA [88] provides the guidance for ensuring the cloud security and provides the certification for same in order to promote the security enabled delivery of cloud services. A Cloud Controls Matrix (CCM) has also been published by CSA that provides the description of key security control which can be used to assess the services of cloud providers. This is very useful document that ensures the effective implementation of the cloud security governance. CCM provides guidance in 16 domains of security including identity and access management, application security. Key management, mobile security and data center operations. The 16 domains primarily focus in three areas of cloud computing: Architecture, Governance and Operation in Cloud Computing. A baseline is set by the CMM for helping organizations to achieve best cyber security strategies. Customers can use CCM metrics to compare various cloud service providers. By following CCM, organizations are preparing themselves to follow other standards such as HIPAA, NIST, ISO 27001, HIPAA, etc.

Security frameworks explained above such as ITIL, ISO/IEC 20000, and ISO/IEC 27001/27002 security frameworks focus on:

- Ensuring that current security policies are according to the need.
- Applying the security baseline in all IT operations.
- Ensuring that all the services are secure from cyber threats.

Incorporating these frameworks in the organization impact the organizational growth and reduces the risk from outsider threats.

2.4 CSA Cloud Reference Model

The cloud security reference model [89] addresses the relationships of different cloud stakeholder classes. As per the security control and concerns, each of them is placed in the architecture. The following need to be known for avoiding confusions in services:

- The term how cloud services are deployed is completely related with where the cloud services are provided. The description of public or private cloud can be done as external of internal.
- It is very important to understand the security boundaries of the organization who should have to follow a well-demarcate perimeter as security perimeter. Cloud services are consumed in the manner of the location of security perimeter of the organization.
- The deployment and consumption of cloud is not only related with the location (internal or external) of cloud services but also related with physical location of resources and who consumes the resources who is responsible for security, governance, policies and standards.

In addition, the location of assets, the security risks are also associated with followings:

- Who is responsible for managing assets and how assets will be managed.
- Which type of assets, information and resources are being managed?
- What are the controls being selected and how they are integrated?
- What are compliance issues associated with services?

Infact ISO/IEC 27002 says that the introduction of external third parties or services should not affect the organization's information processing facilities. There is a difference in the responsibilities and methods for securing different cloud models. It raises various security challenges with customers. Cloud customer can only be aware about the type of security controls and at what level they are implemented if the cloud service provider (CSP) share such information with customers. The customers can be tremendously misguided for making risk management decisions without proper awareness about the security controls. CSA cloud security architecture maps the cloud model with the security controls as shown in Figure 2.4 [89]. Once the mapping is done, it becomes easier to determine what needs to be performed to feed back the risk assessment framework. It helps in deciding how the risk should be addressed,

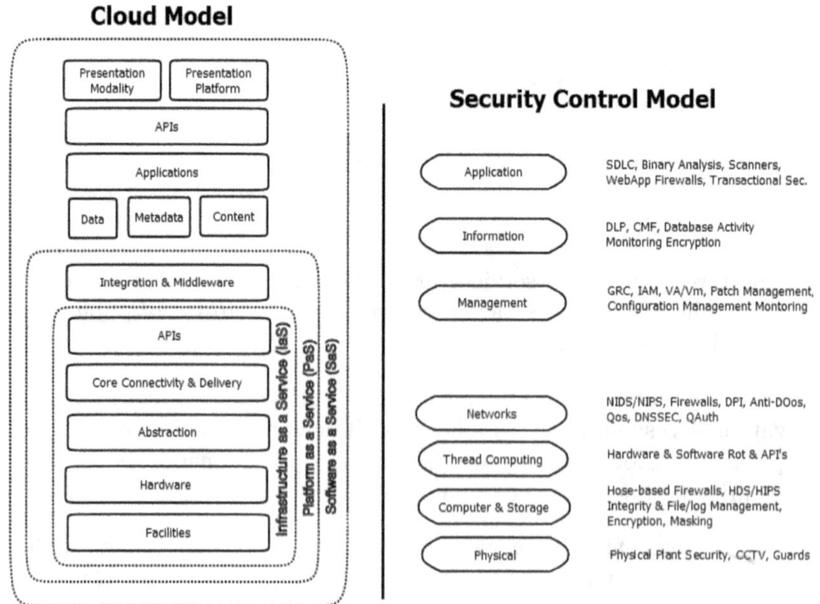

FIGURE 2.4: CSA cloud security reference architecture.

accepted, transferred and mitigated. The gap analysis maps the cloud services and classifies them against architecture model. As an output of this gap analysis, general "security" posture can be determined and can be related to the asset's assurance and protection requirement. It then becomes possible to map the security architecture with regulatory, business and other compliance requirements.

The security controls in IoT are quite similar with the security controls in IT environment. However, because of the various types of service models, operational models and technologies used by cloud computing, there are different risks associated to an organization in cloud environment than traditional IT environment. The security controls are implemented at various layers such as physical security layer, network infrastructure security layer, IT system security layer and information/application security layer. In addition, security controls are implemented at people and process level like assigning different roles and responsibilities and change management, respectively.

The security responsibilities vary for both service provider and service consumer for each of the cloud service models. For example, let's consider the example of Amazon's AWS EC2 service which is IaaS service offered by Amazon. With respect to this service, the service provider is responsible for the security controls such as securing the hypervisor (virtualization-layer), physical security and environmental security, etc. The service consumer will deal with the security controls such as instance security such as security of applications, operating system and consumer's data. However, for SaaS service models such as Salesforce.com's customer resource management (CRM) which deals with the entire stack. The provider not deals with security control of environmental and physical security but also deals with security controls for applications, data and infrastructure. This reduces many of the responsibilities of cloud customer related to directly handling the security controls.

Presently, there is no provision for the consumers to understand for what are their responsibilities for ensuring security. However, CSAs are making efforts to define standards in cloud audits. The cost efficiencies supported by providers are one of the major focuses of attraction of cloud service provider. The efficiencies are provided by making the services flexible so that a large number of customers can takes its benefits. It is not fortunate that solutions integrated with the security are not perceived flexible. The rigidity often comes because of the abstraction in the infrastructure and lack in the visibility which often makes it difficult to integrate the security controls specially at the network layer.

In SaaS environment, the negotiation of the security controls with their scope is done in service contracts. The privacy, compliance and service levels issues are handled legally in service contracts. However, in IaaS services, the key responsibility of securing the underline infrastructure is of service providers and the remaining cloud stack is the responsibility of the consumers. However, In PaaS services, the provider is responsible for securing the platform. The consumers are responsible for developing the applications in a secure

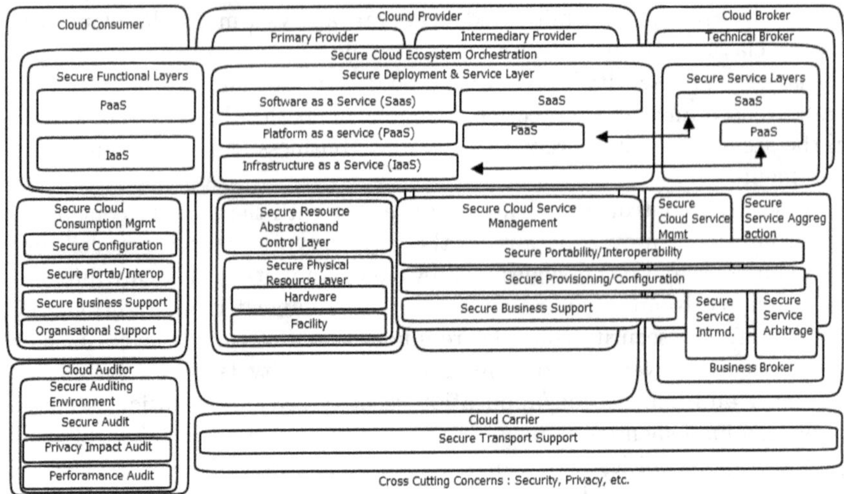

FIGURE 2.5: NIST cloud security reference architecture.

way including the security of the developed applications. It is very crucial to understand the difference between the security control with respect to each service model in order to deal with the risks associated with organizational operations.

2.5 NIST Cloud Reference Model

The formal model of cloud computing security reference architecture (NIST SRA) has been derived from NIST reference architecture (NIST RA) as shown in Figure 2.5 [90] which depicts various architectural components. The architecture depicts that the components identified by NIST RA, should be safe. The ultimate aim of cloud customers is to find out the security controls which best fits the security needs of customers. The ordered list of all the architectural components and sub-components are described as follows:

2.5.1 Architectural components of consumer

The cloud customer's responsibilities were not explained in earlier architecture NIST RA. However, NIST SRA provides the detailed description of the cloud customer's architectural components. The architectural components of consumers are described with their sub components on the following pages.

Secure Cloud Consumption Management

This component includes functions which are very important for the operations of services used by consumers. The various sub components are

i Secure Configuration: It is responsible for the secure management of the following areas: rapid provisioning, resource changing, monitoring and reporting, metering, service level agreement management.

- Rapid provisioning: It refers to automatic deployment of cloud services and resources in the cloud systems. The secure provisioning of resources ensures that the requests are coming from the authorized and authenticated sources.
- Resource Changing: It refers to doing the configuration changes, upgrades and resource reassignments/release, etc. in the cloud environment. The secure resource changing ensures that only authenticated and authorized requests for resource are considered.
- Monitoring and Reporting: The virtual resources are monitored and reports are generated in some time intervals. The secure monitoring and reporting ensures that the limited/required use of resources.
- Metering: It refers to the billing facilities provided by service provider to generate bills based on the usage of resources. The secure metering refers that the pricing model should not be compromised by attacks such as EDoS.
- Management of Service Level Agreement: It refers to managing the SLA definitions, monitoring, and policy enforcement. The secure SLA management ensures the visibility of clauses with customers.

ii Secure Portability/Interoperability: It makes sure that data can be moved securely to various cloud environments. The components mapped to this sub-component provide higher flexibility for the security of data which is being transferred to the different cloud providers. The data security and protection scheme are enforced to mitigate the risks. A greater amount of flexibility is provided for securely transferring data/applications over another cloud provider's platform by incorporating security components at this level. However, there are many other challenges other than security while supporting data migration. Providers are working in this direction to support interoperability feature to students via secure channel. The basic requirements for supporting such as feature varies service model to service model.

iii Secure Business Support: It includes services used for running business operations like management of service contracts, payment, business relationships with other cloud actors and many more, as listed briefly:

- Cloud actors such as Broker, Auditor and Carrier are managed at this level along with their business relationship. The interaction between the cloud actors is formally authorized and authenticated as per the security guidelines.
- Business Issues and other problems related to cloud are discussed with the actors identified for organizations' cloud ecosystem. The best security practices are discussed to ensure the business continuity.
- The service contracts are managed to ensure the secure set-up of contracts, secure termination and closing, etc.
- Only those services are procured where security concerns have been addressed properly.
- The payment and invoices are also managed securely by following best security practices to avoid any fraudulent transaction.

This architectural component also includes various other features such as provisioning of identities and credentials to the employee of organization and contractors by applying access control policies and business continuity plans.

iv Secure Organizational Support: It is responsible for covering processes, policies and procedures given by an organization to support the overall secure consumption management of cloud. The compliance management, audit management and the governance risks and compliance, the standards related to technical aspects, policies and standards for information security, etc., are sub-components of this architectural component.

Secure Functional Layer

The components are used to secure the cloud functional layer and rely on the particular cloud service model used. The modules of this layer are:

First is SaaS cloud ecosystem, in which cloud consumer only has limited administrative control of applications. In another way, it can be said that minimal access control in cloud and security components are provided. Second is PaaS cloud ecosystem, in which cloud consumer has the control on applications. In addition, some control is also given for the hosting runtime environment. However, there is no or limited access to the infrastructure such as access to network, server and storage.

Lastly, third is Iaas cloud, in which cloud consumer has some control in the provisioned infrastructure mainly, virtual machine, middleware and guest OS. The access to virtual computing resources offers more privileges to the users than other service models. However, access to host or physical server and hypervisor is still not granted. For example, a cloud customer can implement a security system such as firewall to secure the infrastructure allocated to it.

The access control policies are specific to a service model. The details about access control, rules, policies and standard is clearly specified in the SLA.

2.5.2 Architectural components of CSP

Various activities are carried out by the service provider such as coordination, management of the resources, etc. A proper ecosystem need to be managed for the provisioning of secure cloud services at various levels such as SaaS, PaaS, and IaaS. The sub-components are as follows:

Secure Cloud Ecosystem Orchestration

Cloud provider performs the functions of orchestration of services, management of services, privacy and security aspects. By NIST, it is described as the composition of various design components that supports which help the provider in coordinating and arranging the cloud components. It has the following components and sub components, as follows:

i Secure service and deployment layer: The requirements of security components depend on the type of service offered to the customers. In case of IaaS cloud, the cloud provider has full control on all the infrastructure resources such as physical server, networking resources and storage. It also executes the management software for accessing the hardware and has full control on it. The cloud consumers are offered the access of virtual resources such as virtual machine, virtual network, etc.

ii In case of PaaS cloud, cloud service provider handles all the computing resources and facilitates the middleware and runtime execution environment. The PaaS ecosystem offers all development tools and/or infrastructure to the customers. PaaS customers have certain control the development tools and deployment infrastructure. In case of SaaS cloud, the provider has complete control on full stack and handles the maintenance, configuration and updation of applications as well. Service provider provisions the services to customers in SaaS ecosystem for effective usage. Customers have limited control in the application only.

iii Secure abstraction of resources and control layer: It represents the security components which a CSP would implement in order to provision the safe access to the physical components. It is the responsibility of the service provider to ensure that proper security processes and components are in place to enable the legitimate access to the resources. The data owned by one tenant should be safe from other tenants sharing the resources. The data of one user should not be accessed by other users. The non-privileged users should not be allowed to access the service management related functions.

iv Secure physical resource layer: The physical resources such as Memory, CPU and networking resources (firewall, router) and storage devices (Hard disks, etc.), should also be secured from the unauthorized access. This layer facilitates such security components which ensure same. The other resources such as ventilation, heating, communication and power are also included at this layer.

v Secure cloud service management: It has been described with the perspective of requirements as follows:

a Secure provisioning and configuration : It provides the following functionalities:

- Rapid provisioning: The services are provisioned and deployed automatically on receiving the service demand from the customer.
- Resource updation: It refers to upgrading and maintaining resources as per the modified security policy.
- Resource Monitoring and Reporting: Cloud virtual resources and cloud events are monitored and security reports are generated which also comprises of resource usage statistics.
- Metering: The billing facility is provided with some abstraction based on the service type (processing, storage, etc.)

b Secure interoperability and portability: It provides the assurance that the customer's data can be easily transferred securely to various cloud ecosystems as per the requirements of security as identified by SLA. The maintenance time and downtime should be fixed at minimum level.

c Secure business support: It comprises of business related services which supports the security processes of customer's and provider's. For example, contract management, customer management, inventory management and reporting and auditing, etc.

2.5.3 Architectural components of broker

The services defined for cloud broker can be classified into following categories:

i Secure Service Aggregation: The multiple services are combined, forming one or more service platforms. The secure movement of data and its integration is also the responsibility of the broker. The various key sub functions are: secure configuration and provisioning and secure interoperability and portability.

ii Secure Management of Cloud Service: Various security functions which are required for the secure management of various services are included here. Sub components derived from the activities of this component includes secure portability for supporting secure transportation of services, secure provisioning of resources, tools, etc. and secure business support.

iii Secure Service Intermediation: It addresses the brokers responsibilities for ensuring the capabilities added to existing cloud services.

iv Secure Service Arbitrage: The secure service aggregation and service arbitrage are very similar in various aspects. However, in later, the combined services by broker are not fixed and are flexible. They can be assigned to multiple vendors dynamically. All the functionalities of the broker has been specified and explained above. Broker is supposed to perform all its functions, considering in mind the security requirements specified in SLA.

2.5.4 Architectural components of carrier

The transport and connectivity of the services is provided by the actor. The secure management of services is also performed by the cloud carrier in order to ensure that customer satisfaction about the service delivery is maintained. The cloud carrier provides the way via which cloud customer can directly contact the broker or provider. The complex details in order to maintain the connectivity is concealed from the consumer. Although carrier plays an vital role and makes it possible to have a transportation of services between provider and consumer without having a cloud carrier. Cloud service provider ensures that a secure channel should be provided to cloud customers to meet the SLA requirements which are signed between them. In addition, service management functions are also performed by the cloud carrier to ensure the effective service delivery to the customers.

2.5.5 Architectural components of auditor

Cloud auditor is responsible for performing the independent assessments of the services. It ensures that system operations, security policies, privacy functions, etc. are properly functioning without compromising any service quality. The secure auditing environment includes various mechanisms such as security controls, secure archival, secure storage, data location, metering, SLAs and privacy, etc. It requires the following security mechanisms in place

- Secure Archival: The archival of all the audit records for the business and legal processes should be secure. The requirements for archival and its implementations should be available to auditors.

- Security Component and Related Control: The detailed information about the various security controls and security components should also be available to auditors.

- Secure Storage: The evidences can be collected from concerned authorities and stored in cloud safely for future purpose. Therefore, the secure mechanisms such as obfuscation and encryption information should be provided to the auditors.

- Data Location: Some jurisdictional rules might be required to be applied in data during the process of assessment. It requires the availability of information about data location to auditors.

- Metering: The metering/billing information need to be given to auditors for the performance audit. The access should be granted in secure manner.
- SLAs: Auditors are also required to have a secure access to the all the agreements which are implemented between the parties.
- Privacy: The privacy assessment by auditors also requires the secure access to the information about the configuration information and system security which are implemented by organization for protecting the client's data.

The security controls along with the security components are independent of the cloud service model and available to all the cloud auditors.

2.6 Conclusion

Cloud Security plays an important role in building a trust between cloud service provider and cloud consumers. It provides tools and technologies to protect the infrastructure, applications and data. Various security aspects for cloud are same as the on premise organization. However, the addition of virtualization layer has opened the doors for attackers and hence requirement for specialized security techniques to deal with cloud specific threats. Various cloud security concepts have been discussed such as multi-tenancy, virtualization, data outsourcing, trust management and meta security, etc. Afterward standards for cloud security have been described such as Information Technology Infrastructure Library (ITIL), ISO/IEC 20000, Statement on Standards for Attestation Engagement (SSAE), Cloud Controls Matrix and Cloud Security Alliance (CSA), etc. A cloud service provider has to follow the security standards in order to maintain the security of organization and customer' data. At the end, some of the important cloud security reference architecture (such as NIST, CSA) have been discussed to provide an overview about security architectures in cloud.

2.7 Questions

Fill in the blanks

1. Which of the followings are cloud security standards

   i ITIL

ii COBIT
iii ISO/IEC 20000
iv All of above

2. Mark the appropriate sentences. The security risks are associated with

   i Who is responsible for managing assets and how assets will be managed?
   ii Which type of assets, information and resources are being managed?
   iii What are the controls being selected and how they are integrated?
   iv What are compliance issues associated with services?

   Choose the correct option for answering the question 2.
   (a) i (b) iii
   (c) ii (d) all mentioned above

3. Type-1 hypervisor(s) is/are

   i Xen
   ii VMware ESX/ESXi
   iii Hyper-V
   iv All of above

4. Which are the key vulnerabilities in cloud environment

   i Lack of physical control
   ii Under-provisioning of bandwidth
   iii Pricing model of cloud
   iv Insecure Browser and APIs

   Mark the correct option for answering the question 2.
   (a) i & ii (b) iii & iv
   (c) ii & iii (d) all mentioned

5. Mark the incorrect statement. The attacks which exploit protocol vulnerabilities and affect pricing model of cloud are

   i RIP attack
   ii DNS Provisioning attack
   iii Flooding attack
   iv Economic denial & sustainability attack

Short-Answer Questions

1. Explain various functionalities of auditor with associated security concern.
2. What do you mean by cloud security standards. Explain COBIT.
3. What is ITIL standard? How does ITIL help in meeting the security requirements?

Long-Answer Questions

1. Define Cloud security. Discuss all the architectural components of NIST cloud security reference architecture with suitable diagram in detail.
2. Explain various cloud vulnerabilities. Also discuss the need of cloud security. How cloud security is different than traditional security environment.

Chapter 3

Cloud Security and Privacy Issues

3.1 Introduction

The emergence of various cloud-based services has opened good opportunities in various domains such as Internet of Things (IoT), Smart Grid, Healthcare, Banking, and IT. However, security is one of the crucial aspects in the cloud computing, which has been studied in detail by cloud service adopters, researchers, and security professionals. Cloud offers various good features for the better utilization of the resources. Many of such features have been discussed in previous chapters. However, various features such as multi-tenancy and online access to data and applications from anytime and anywhere expose some serious threats as well.

For example, multi-tenancy could be misused by some of the cloud tenants to cause harm to the shared cloud resources and breach the security of co-located VMs. Moreover, the availability of services can also become threat to the cloud infrastructure as services are provisioned in online mode. Advanced attacks can eavesdrop the network connections and can gain access to the information being shared between sender and receiver [60]. Moreover, the data stored in the cloud storage servers can also be exposed to third party organizations intentionally for gaining some financial benefits. Since the data is stored in the shared storage resources. If proper isolation of virtual storage volumes is not maintained in the physical storage, then it will be easy for an attacker to access the data of other customers.

Furthermore, the vulnerabilities present in any component of the cloud infrastructure such as controller server or computer server, network server, hypervisor, virtual machine and user applications, etc., impose a direct threat to the security and privacy of services. Some of the other vulnerabilities which can hinder the security and privacy are: insecure live migration of the customer's data, random selection of cloud service provider, in secure application and browser APIs at provider end, network vulnerabilities and insecure encryption of data, etc. [91]. The vulnerabilities and threats present in cloud, make it very difficult to develop a comprehensive security model that can cover all possible vulnerabilities.

Due to some security reasons, cloud service providers do not allow the cloud customer to impose their own security model at the cloud network, or

DOI: 10.1201/9781003004486-3

integrate them with management services or API, etc. It is the responsibility of the provider to secure each layer of cloud infrastructure. The customers can however install some security tool inside their own VM for the analysis of their own VM's activity. If the VM is subverted, the security tool inside VM will also be subverted. Therefore, security and privacy are major concerns for cloud service vendors and end users before migrating their applications, data or any services in cloud.

Security refers to the protection of cloud infrastructure along with associated data and applications from the unauthorized threats and attacks. The protection of cloud resources is important to ensure the confidentiality, integrity and availability (CIA) in cloud environment. For example, the unauthorized users should not be allowed to access and modify the cloud resources. The services should be available anytime from anywhere and service outage should be as minimum as possible. The user's data could be very sensitive and protection of data from the attackers is the responsibility of the cloud service provider.

Privacy is one of the important rights of a person to not to share his private data to public. Private data refers to the personal sensitive information of person which he or she may want to share to restricted number of people or with no one. In the context of cloud computing, privacy can be defined as "the obligations and rights of an organization and individuals with respect to the gathering, usage, retention, disposal and disclosure of private information" [92]. The privacy concerns are also different for each cloud models. The security concerns for public cloud are much higher than private clouds or hybrid clouds. As publically cloud can be accessed by anyone over the web and hence more prone to security attacks. However, in private cloud, the resources are shared mainly by organizational people and attacking risk is less due to having restricted and controlled access to the resources.

Although security and privacy are defined in different ways. However, they are highly related with each other. The security is a much broader term which is used to ensure various security aspects, mainly confidentiality (unauthorized disclosure) , integrity (illegitimate modification of data) and availability (smooth access to services) of services. Whereas privacy is mainly concerned with the ability of an individual to seclude their information from public and share with selective people based on personal choice. Various privacy preserving security frameworks [93, 94] have been introduced to preserve the privacy of individuals and at the same time meeting other security goals of cloud. Therefore, security is not privacy but techniques used to ensure that privacy and others security factors such as CIA, are persevered in cloud.

However, there exist less transparency when we talk about attacking incidents and corresponding security measures taken by vendors, in cloud environment. If some security incident occurs, customers may not be aware about same. In fact, the detailed report about the incidents such as vulnerabilities exploited, malware details and any system damage, etc. is also not shared with the customers. For example, there exist some backdoor attacks which are launched by attackers to again access to the VM instances running in cloud. The detection and notification of such backdoor attacks is essential to have in the primacy stage. However, as long as no such backdoor is reported by provider, customers are still blindly trusting the provider's security and assuming that no such attack has happened. Although the security and privacy related aspects and services are usually highlighted in the Service Level Agreement (SLA). Various popular cloud companies such as Google, Amazon, Salesforce, etc. are dependent on detailed SLA to ensure the customers that strong security is provided. However, there exists no one standard for designing the SLAs.

Various cloud features also expose threat to the security and privacy of the user's personal information and cloud infrastructure. Handling and addressing all possible security issues at all possible layer in cloud is still a challenging task. Gradually, the enterprises or cloud service providers are expanding their cloud services and hence the attack vector associated with various services is also expanding. Each of the organization is interested in incorporating a security framework for running the services in a protected environment. Moreover, the security and privacy of the user's data is as important as the infrastructure security. Each organization has to take care of choosing a secure cloud computing platform before migrating their services and data to the remote cloud servers.

3.2 Cloud Security Goals/Concepts

Cloud security is one of the essential need of today' eras as most of the organizations are moving toward the adoption of cloud service platforms. There is a strong requirement of having the proper controls, security policies, defensive mechanisms, and procedures in place so as to protect complete the eco-system. In this section, we will understand, various cloud security goals that every organization wants to achieve.

3.2.1 Confidentiality

The confidentiality of data is a crucial issue when extremely sensitive data is outsourced to the cloud system. It keeps the data secret from the users in the cloud and the confidential data must not be accessible to an unauthorized entity. To achieve confidentiality, mechanisms such as cryptography and isolation has been adopted by cloud vendors. Encryption mechanisms such as triple Data Encryption Standard (DES) or Rivest, Shamir, Adleman (RSA) are used to gain confidentiality but key management or key distribution is the big issue. Some examples of threats to confidentiality are insider user threats such as malicious cloud provider users. Malicious cloud user and malicious

FIGURE 3.1: Cloud security goals.

third-party user. The attacks by external attackers such as the attack on application or infrastructure by remote software or hardware. Data leakage is another threat to the confidentiality.

3.2.2 Integrity

Data integrity is the basic task that verifies the data and it provides the guarantee for the exactness and quality of the data. It is important as the cloud provides various services such as SaaS, PaaS, etc. The cloud services demands have been increasing day-to-day, hence cloud service providers may require increase storage. So, there are chances of data corruption or loss or maybe the cause of failure of nodes, physical devices, or disk. The data integrity is preserved in the cloud environment by various means so that data are not be altered by an unauthorized entity. To avoid data corruption or crash in the cloud, so watching the data integrity is very essential. As cloud-based environments are distributed so it is harder to obtain integrity as compared to the centralized environment. Examples of threats to integrity like user access, data segregation, and data quality.

3.2.3 Availability

In the cloud-based system that includes application and infrastructure, the goal of availability is to provide the services to its users from anywhere and at any time. But some circumstances occur in which the availability of data cannot be sure. There may be unavoidable circumstances such as natural tragedies, hence it is essential to know the data can be used, authenticated, or restored by the data users. The cloud users must know about the security actions that are to be taken by the CSP and must read the Service Level Agreement. The availability of cloud services is obtained by using fault-tolerant systems in the cloud environment that can tolerate the failure of the server or cloud. Redundancy and hardening are two mechanisms that can be applied to increase the availability the services in the cloud-based system. Threats to availability such as the denial of services like network DNS, data, and application. Liu [95] discussed a novel cloud Denial of Service form. The effect is normally reliant on existing, processing capacity, memory, and bandwidth in flooding attacks.

3.2.4 Authentication

Authentication is the method of creating assurance in the identities of the user. Authentication guarantee levels must be suitable for the application sensitivity and information resources accessed. An identity management schemes can be used to authenticate users and cloud services using credentials. A big challenge related to Identity Management (IDM) in the cloud is interoperability limitations. Password-based authentication techniques have a genetic drawback and have important risks. The IDM must secure sensitive and private data concerning to users. As cloud service providers have been increasing day by day they support a standard SAML that is used to authenticate the users before administering the data and application access. SAML offers mechanisms that exchange information among cooperating concerns. The request and reply messages of SAML are mapped on SOAP that uses the XML format. Chow et al. [96] discussed that authentication is needed before offering access to Software as a Service application is beneficial due to centralized monitoring.

3.2.5 Authorization

The sensitive information and services of the users can be accessed by unauthorized users. To restrict data access authorization must be used. The identity management system should be employed. Authorization is used to control the access of data. Authorization is the method that allows a system to regulate access level to a specific authenticated user. There are benefits of centralized access control and alleviate many security and management actions. Though that cannot be desirable in a case populated with data mix-up, that can be happened in the future [96]. This is risky to gain access te sensitive information when authorizing third party service. Grobauer et al. [91] recognized faulty or unsatisfactory authorization tests as expected vulnerable vectors.

3.2.6 Auditing

Auditing is the monitoring task to know what is going on in the cloud-based system. An additional layer can work for audibility in the virtual machine to monitoring the system. As it can monitor the complete access duration, hence it is safer than that is made in software or applications. Audit approaches investigate service conditions, monitor malware, accesses, and other actions, and record logs with an exhaustive explanation of what occurs are appropriate. The audibility simplifies the method of recognizing the authorized party legal action situation that can be important to the cloud stakeholders. Auditability contains in acting tests series to discover if all suitable implementations conform. In cloud environments, the other layer above virtualized guest operating systems will also agree that [97].

3.2.7 Access control

To minimize the security risk, access control features can be used that maintain the control on access to the resources. Access control is part of identity management. The eXtensible Access Control Markup Language (XACML) is the standard that can be used for resource access control in the cloud environment. XACML focus on techniques to take authorization decisions that complement the focus of the SAML on the resources for carrying the decisions of the authorization and authentication among cooperating domains. The protocols or transport methods are not provided by the XACML and it also does not describe the method of validating the credential of the users. The transmission of the message among XACML units is vulnerable to attacks that can be performed by malicious third parties such as replay, unauthorized expose, loss, and alteration attacks.

3.3 Cloud Security Issues

The rapid growth of cloud services and increased demand in the organizations, have raised various security concerns. If there exist any flaw in the implementation of any cloud service, it could be a big threat which could raise serious security issues. The migration or transitioning of services to public cloud environment also causes the transfer of responsibilities to service provider to protect the data, applications, information and infrastructure, etc. It removes the direct control on the cloud resources and management operations from the customer side. This makes the customers highly dependent on the cloud vendor for the smooth functioning of services. Various security critical operations like incident response, auto-updation, continuous monitoring of services, etc. are now performed by cloud vendors. Hence, cloud customers have to trust on the clouds vendors. However, it has been reported by various organizations such ENISA [49] that insider threats can also cause major harm to the cloud resources. The 65% of insider threats can harm the reputation of an organization and can affect the finances. Since, the customer's data is in remote servers, which is sharing its resources with other customers, the data breach can also occur from the outside personnel. Although there exist various protection laws and regulations which provides the guidelines for joint

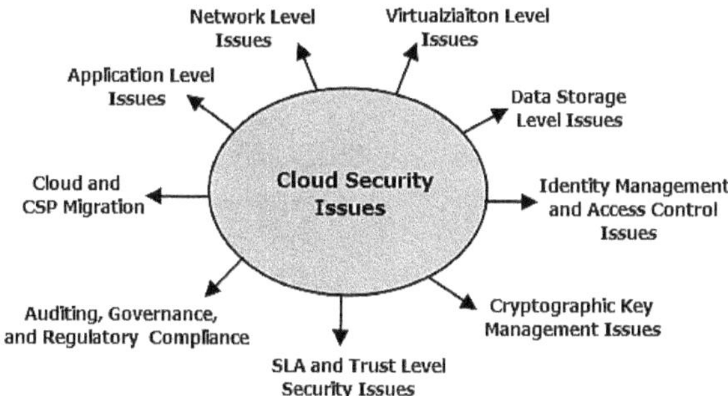

FIGURE 3.2: Cloud security issues in cloud.

responsibilities so that proper cooperation between cloud vendors and customers can be achieved.

However, due to lack of physical contact with the servers and cloud personnel, its again challenging to enforce such guidelines. There are various cloud security issues at each of the layer of cloud environment. These are application level, virtualization level, network level, data storage level, cryptographic key management level, identity management, and role-based access control, SLA and trust level, auditing, governance and regulatory compliance and cloud and CSP migration level security. Application level security issues are concerned with the vulnerabilities and threats present in the web applications, web browsers and application layer protocols, etc. Network layer security issues are concerned with the security vulnerabilities present with network layer protocols, network servers, networking applications and services and any attacking study that is targeted through network.

The virtualization layer security issues are more concerned with the threats and vulnerabilities associated with the virtual machines and hypervisor. The data storage level security issues are concerned with the storage layer attacks and vulnerabilities that can be exploited by attackers. The cryptographic key management is more or less concerned with the key management schemes and their limitations. The identity management and role-based control issues are mainly associated with vulnerabilities in customers' identity schemes and granting the resources to them. The SLA and trust level issues talks about the flaws present in the implementation of same. The auditing and regulatory policies are also major security threat as improper implementation of same, can cause severe security breach. There are various security issues associated with live migration of cloud services as well. Let us discuss each of the security issue one by one. These issues are also shown in Figure 3.2.

3.3.1 Application level security issues

Application level security issues are concerned with the security of web applications running in the cloud to provide cloud services. The SaaS application has to be managed over the web (using a browser). The web application security is tightly coupled with the security of web browsers. A web browser is a platform independent program, used to access the cloud services (SaaS), web 2.0 or web pages. A web browser uses SSL/TLS protocol for secure transmission. The security loop holes in the web applications create the vulnerabilities in the SaaS applications. The web applications are prone to a number of threats such as cross-site scripting (XSS), SQL injection attack, broken authentication, insecure transport layer protection, cross-site request forgery (CSRF), etc.

The security at the application level states the use of software and physical resources for protecting applications such that the adversary cannot gain control over applications. Application-level security issues are concerned with the security of web applications running in the cloud to provide cloud services. The SaaS application has to be managed over the web (using a browser). As software-as-a-service and web applications are tightly-coupled with offering services, cloud services availability and protection rely on Web browsers and APIs security. A Web browser is a platform-free client program by which clients can access the SaaS and web applications. The protocols TSL/SSL can be used to authenticate and protected data transfer. It uses SSL/TLS protocols for secure transmission and authentication of data. The web applications are prone to several threats such as cross-site scripting (XSS), SQL injection attack, broken authentication, insecure transport layer protection, cross-site request forgery (CSRF), etc. Hence, the adversary can breach the security of cloud applications when they target Cloud authentication based on the browser. Any adversary can gain access to XML tokens that are authentication-related passes of another customer that is helpful to get access services of the target. The XML encryption and XML signature is the useful mechanism to improve the security of the browser [64]. Though, the XML Signature Wrapping attack allows the adversary to modify the content of the signed portion and invalidating the signature is not included. Some AWS accounts may be taken over due to the cross-site XSS scripting vulnerabilities that may be the cause of the XML signature wrapping attack.

3.3.2 Network level security issues

Network level security issues are concerned with the security of the cloud network. One of the key issues at network level is unavailability of services. Denial of Service (DoS) and Distributed Denial of Service (DDoS) are main threats to service unavailability. These attacks cause inconvenience to customers and prevent their access to the cloud services. HTTP-based and XML-based DDoS, are called as Economic Denial of Sustainability (EDoS) which affect the pricing model of cloud [98]. The key security issues at the network-level are authorization, authentication, intrusion detection, vulnerability assessment, session hijacking, etc. Some common attacks at network-layer are sniffing, scanning, IP/MAC spoofing, and DNS poisoning, etc.

The authorized clients of cloud who may be inside network adversary and they can gain access the resources of the other customers. The internals is privileged and they have more information than the external adversary. This information is important to know about the network, security approaches, and resources. Hence, this is convenient for an internal to perform the attacks than external adversaries. The key security problems at the network level are backdoor attacks, Internet protocol vulnerabilities, session hijacking. Many cloud service providers such as Azure, Amazon, etc. use a firewall to cope up with few challenges at the network level, but it cannot be helpful for inside attacks. Some challenges can be solved using integration with network-based IDS. Though, a network-based IDS must be set up for sensing not only external but also internal intrusions. This must also be proficient in sensing intrusions from encrypted traffic.

3.3.3 Virtualization level security issues

Virtualization level security issues are concerned with security of virtualization layer. The major vulnerability is the multi-tenancy in which multiple tenants share and utilize the cloud platform [60]. As the number of TVMs running above Hypervisor increase, the security issues with the new TVMs also increases. Maintaining the security policies of all TVMs is challenging task. Malicious code running inside a TVM may try to gain root privilege of the Hypervisor with an intention to take full access of the system. Security of TVMs is one of the crucial concerns in the cloud environment. Once Hypervisor is compromised, all TVMs running on it will be under the control of the attacker [99]. Infact, an improperly configured Hypervisor can fail to provide proper isolation among TVMs, leading to disclosure of the tenants' sensitive data.

When more virtual machines are added to the hypervisor, the security concern with new VMs increase as this is impossible to keep up with all virtual machines. Therefore, it is hard to maintain the security of those VMs. There may occur that a guest VM attempts to execute a malicious code on the host to gain full control of the system so that other VMs access could be blocked. The physical infrastructure is shared among multiple clients and if one user is malicious, then he can be a threat to other users who are sharing the same infrastructure. If an attacker can gain control over the hypervisor, he can modify any guest operating system and gain control over the contents moving through the hypervisor. The attackers can exploit the vulnerabilities of the hypervisor and get control over the virtual machines. Some examples of attacks are Subvit, DKSM, and Bluepill that are still an open challenge.

3.3.4 Data security

Data security is vastly an open research area. Data can be in transit (communicated via network channels) or in rest (stored in data centers). The vulnerabilities in the network protocols and/or poor encryption directly affect the confidentiality and integrity of data. The data stored in the servers needs to be physically and logically segregated and have control policies. A few years back, Amazon reported that its Elastic Block Store (EBS) volumes were trapped which affected its EC2 instances [100]. During data-in-transit, the attack in the network will affect data integrity and confidentiality. The major risks in the case of data-in-transit can be network protocols and bad encryption approaches. Data linage refers to trace the path of the data and this is essential for auditing in the cloud. This is a critical task for the provision of data lineage. As the data flow in a virtualized environment within the Cloud is no longer linear, it complicates the task of mapping the data flow to provide data integrity. Managing data integrity is a critical issue in the Cloud, because of the shared environment. Data-at-rest refers to data that is kept in cloud storage that requires the policies of the logical, physical, and access control.

The major issue in data-at-rest in the cloud is losing control due to the data accessing by unauthorized users in a shared environment. In-built encryption schemes in storage are not able to avert unauthorized access as an attacker can steal keys used for encryption/decryption. A lockbox method in which the real keys are kept in a lockbox and there is another key for the lockbox that can be used for the mentioned scenario but with the requirement of lockbox key, it can cause key management the issue. Data remanence refers to data which is left out after transfer or removal of VM. Data recovery is preferred when data is lost because of some accidental damage. Data segregation is the organization of the data of various users residing in the same location. Ensuring the isolation between the user's data is an important security concern. Data integrity ensures that there is no illegitimate modification in the user's data [101]. Data deduplication is an approach for removing duplicate copies of data. Secure data deduplication is a major research concern [102]. Data recovery is another biggest challenging issue. There is the possibility for natural tragedy or accidental harm to the storage devices and due to these reasons, data can be destroyed and data availability can be at risk. Data location, finding the data location is critical in the cloud environment, as data of the user are placed dynamically from one country to another. This raises security and data privacy risk as the data possessor can lose control of his data. In data moving or data removing, data left out, known as data Remanence. Due to it, there is less security threat such as the expose of critical information.

3.3.5 Identity management and access control

Identity management and access control issues are also important security concern. Identity management (IDM) deals with identifying the entities in cloud and controlling their access to resources. Information privacy and sensitivity are highly dependent on the IDM policies and mechanisms in cloud. Identity authentication and verifying them are the features of identity management. So, avoiding information accessing from unauthorized access is a big concern in the cloud. Identify management is a wide administrative area that has the responsibilities to recognize the individual entities, cloud entity, managing resource access as per predefined policies [103].

As the customers' credentials are transmitted via internet, it imposes a great risk to user's sensitive data. The issue is addressed by providing the support for federation protocols such as Service Provisioning Markup Language (SPML) or Security Assertion Markup Language (SAML) [104] to some extent. SAML supports both authentication and authorization. Some other protocols are created after SAML such as OpenID and OAuth2. OAuth2 is an open standard for authorization and OpenID is an open standard for authentication. The cloud-based IDM are prone to serious threats such as brute-force attack, cookie replay attacks, eavesdropping attack, denial of service attack and data tampering attack, etc. There is a need to design strong security measures for IDM systems. There is a need of providing fine-grained access control mechanisms for controlling access to user's data. For example, Google App uses eXtensible Access Control Markup Language (XACML) for authorization and access control. Mon et al. [105] combined the Attribute-based Access Control with Role-based Access Control (RBAC) to ensure the privacy and security of user's data.

3.3.6 Improper cryptographic keys management

Improper cryptographic keys management leads to failure of cloud security measures [106]. The cryptographic approaches such as cryptographic hash function, digital signature and message authentication code, etc. are used to authenticate the VM templates in cloud. They may prone to the bootstrapping problem and hence, requires a strong security analysis. The key security requirements for key management systems must be ensured. Some of them are discussed as follows. The key management commands and data should be secure from spoofing and illegitimate modification. The third party who does key management should be authentic. All the secret and private keys should also be protected from disclosure. The cryptographic mechanism employed for protecting keys should be strong enough and robust from attacks.

There are a variety of security algorithms that can be used to confirm data privacy and confidentiality from service providers. An encryption approach provides adequately robust security. ABE (Attribute-based Encryption) is a recently designed public key cryptographic approach that works in a one-to-many manner, also known as fuzzy encryption. Public key encryption approaches save encrypted data on third-party servers and decryption keys are distributed to authorized clients. Though, there are various limitations in it, such as this is hard the private keys distribution efficiently to the valid customers, scalability and flexibility problem and It is necessary for data possessors to be online when data is encrypted or re-encrypted data, or private keys are distributed. The Attribute-based Encryption approach reduces the mentioned confines by minimizing the internet overhead of communication and growing scalability, and fine-grained access control and flexibility for huge-scale systems.

3.3.7 Service level agreement (SLA)

Service level agreement (SLA) and trust level security is another important concern. The customers lose their control over data and programs which are outsourced to cloud servers. Cloud service providers limit the visibility of data location, network and system monitoring to customers which generate the trust issues with service provider. It is very difficult to assure trust in cloud environment. However, the use of signature techniques and advanced cryptographic techniques can be used to deal with the trust issues to some extent. SLA is another way to deal with the trust issues to certain limit. An SLA is signed at the time of registration, describing the minimum performance criteria a CSP should meet when delivering services. If a certain service fails to meet the customers need or quality of service (QoS) do not meet the SLA, cloud customers can lose their trust with the CSP [64].

The clients do not have total control over the cloud resources, but they must ensure the availability, reliability, and resource quality offered by cloud service providers after cloud migration, which is possible with a service level agreement (SLA). With SLA, the clients can use cloud services securely, hence it needs reputation administration. SLA comprises performance, coarseness, clarity vs. complexity, and tradeoffs to fend client requirements and expectations. The cutting-edge SLA schemes rely on clients' feedback concerning main issues such as Cloud interoperability and what is to be migrated. There are various kinds of resources for cloud migration, such as applications related to business, IT organization, deployment of the application. There are various issues that are required to be addressed when business data is migrated. The other challenge is interoperability, in which many clouds that have different methods of client communications with the cloud. Interoperability purposes of identifying the smooth data flow across cloud.

3.3.8 Regular audit and compliances

Regular audit and compliances to manage cloud resources must be done to ensure whether internal and external processes are meeting the customer requirements, regulations and laws. The policies should be monitored regularly. There are some general governance standards that are also applicable to cloud computing environment such as ISO/IEC 38500 – IT Governance [107], Control Objectives for Information and Related Technology (COBIT) [108], Cloud Security Alliance (CSA) Cloud Controls Matrix [109], etc. The law and regulations of different countries are different. Therefore, some of the compliance operate at country-level, or regional-level [64]. Some of the standards are applicable to specific company or data. The Health Insurance Portability and Accountability Act (HIPAA) [110] requires the U.S. health care organization to maintain the confidentiality of protected health information (PHI). Payment Card Industry Data Security Standard (PCI-DSS) [111] defines the minimum security controls to secure the customer data. The Federal Information Security Modernization Act (FISMA) [112] is a compliance framework that enforces the protection of information systems and assets of all federal government agencies and contractors. Sarbanes-Oxley Act (SOX) [113], a federal regulation, provides the standards for all U.S. publicly traded companies to ensure security to all shareholders and public from fraudulent actions. It maintains the information policies and prevent the illegitimate data tampering.

3.3.9 Cloud and CSP migration, SLA and trust level issues

One of the major issues in the cloud environment is trust level issues. As cloud customers have control deficiency on resources, they have to depend on trust schemes agreements in alliance with schemes that offer compensation. In a heterogeneous environment, trust calculation is complex which is measured by a social trust or human. Services may be sub-serving without awareness of the customers. The customers have less visibility of system monitoring and networks that is a big trust challenge. The staff who have authorized access and can be malicious insiders in the organization and attacks could be executed that can influence the privacy and confidentiality of other customer's data and also resources. There can be a trust problem due to public relations lacking. Trust problems can be addressed by offering suitable measures for the visibility of the observing system. There must be means for dealing with the related risks. Access control vulnerability, Cross-site scripting, doubtful configuration, and storage are few examples of threats.

Service Level Agreements (SLA) is an agreement between the service providers and their clients that documents the services provided by the providers and states the service standards. Most SLAs attention to contracts concerning the attempt that will be performed by Software-intensive systems (SIS) providers when the problem takes place. Though, no assurances are stated concerning the service's efficiency for business processes of the client and their business purposes. The issue of availability, unintended resource allocation, deceptive computation, and loss of data are, however, issues that can tamper SLAs [114]. Cloud and CSP Migration, when the users migrate to the cloud, they move their complete setup to the cloud. Where the provider will maintain the computing environment. Though, that is a difficult procedure for several organizations since they had to leave off a specific level of control to the cloud provider. Also, the transfer in itself is a challenge since there are certain aspects in it that the user has to be attentive. When an organization or cloud customer is entering into the cloud or shifting from one CSP to another CSP, the following migrations will be considered: Data (application) migration and Cloud migration. Migration is one of the challenging research areas.

It involves the secure transmission of the tenants' data with strong application and network security measures together with governance compliance. There are many questions that need to be resolved with tenants such as What technology is used in migration? Is the CSP migrating the data with appropriate policies in place? Is the migrated data secure? Is the migration secure from attackers? etc [115].

3.3.10 Hardware-level security issues

The hardware layer is the lowest level in the stack. The hardware or physical level is employed in data centers. At the hardware level, all hardware resources such as physical servers, switches, routers, cooling, and power system are maintained. The hardware layer can also be termed as the physical or server layer. The applications are used by the service providers to observe memory, CPU loads, storage, etc. The users can interact only with the virtualized environment. The adversaries can exploit the hardware layer and can gain physical access to the system. They break the security at the hardware layer and perform attacks on the data integrity and privacy that exist on the secondary storage and main memory. Trusted Platform Module (TPM) may be altered to dump the data of the internal registers and sensitive data can be fetched like the secret key. The adversary obtains a message that is exchanged between TPM and authorized user and after that message can be misused maliciously. Attacks such as side-channel can be performed. The access to the physical resources such as network devices, storage, and processing servers must be constrained physically only authorized persons must be allowed with security authorization to manage the actions. Some hardware-based solutions [116] for cloud focus more on hardware security.

Researchers are working in different domains of security as discussed above to address the security issues. We have considered intrusion detection as one of the key security aspects to detect attacks at different layers in cloud. The security issues are briefly described in Table 3.1.

3.4 Security Requirements for Privacy

Privacy is a more critical issue than security due to dealing with the public. The cloud service provider has the opportunity to examine and escapade large volume of personal data one example is, the service provider could know the number of persons who are suffering from cancer due to showing their interest to search chemotherapy which could be shared to the organization related to the insurance that could use this information to categorize a person as higher-risk for greater premium. We discuss some security requirements below:

TABLE 3.1: Security issues

Security Issues	Sub Category	Threat/Vulnerabilities	Solution
Virtualization level security Issues [117, 101, 91]	Monitoring virtual machines	Untrusted hypervisor element, hypervisor-based rootkits, Hypervisor inspection, isolation, interposition, zero-day attacks, VM escape	Better authentication and authorization. Strong isolation between VMs. Use of secure hypervisor. Monitor activities at the hypervisor.
[118, 119]	VM-level	Side-channel attacks, VM hopping, cross-VM attacks, Malware injection, Covert-channel attacks, VM reset vulnerabilities	
[91]			
[120]	Virtualized networking	Packet sniffing and spoofing, Impact of network security devices on virtual networks, Virtualized communication channels	
[91]	Managing images	VM sprawl, Image stealing and code injection, Large-sized images cryptographic overhead	
	Mobility	Live VM migration MITM attack, VM mobility, VM cloning	
	Malware	VM rollback, Malware escape approaches, Malware expanding to the VMs	
[67] Service Level Agreement	Cost	Service price associated issues,	Cost management based on SLA overcomes these problems and gives an effective solution for controlling the cost of allocated resources based on the hosted application's defined policies.
	Resources	Not giving the particularized quantity of resources,	
	Risk management	Risk computation management to control the resources allocation based on SLA	

TABLE 3.1: Security issues

Security Issues	Sub Category	Threat/Vulnerabilities	Solution
	Waiting time	Customers waiting in line for an extended duration for using service	
	Performance	Service Performance may be affected due to the high workload. The clients will not be fulfilled due to bad performance and the revenue will be affected	
	SLA negotiation	Applications must perform in the care of their consumers and cloud services must be available in SLA negotiation.	
Cloud and CSP Migration, and Trust level issues [121]	Service provider, cloud migration and Trust	Cross-site scripting, access control weaknesses, insecure storage, and insecure configuration	Advanced cryptographic techniques and signature techniques
Application Level Security Issues [64, 122]	Unauthorized Accessing	SQL Injection attack, EDoS, Cross-site scripting, Cookie Poisoning, Google Hacking, Hidden field manipulation, Backdoor and debug options	Check service integrity using a hash function. Web service security. Use secure web browsers and APIs.
Data security	Data Integrity, Confidentiality and Access	The adversary in the network affects the confidentiality and the integrity of data, non-authorized user accesses	lockbox approach

TABLE 3.1: Security issues

Security Issues	Sub Category	Threat/Vulnerabilities	Solution
Identity management [123] and access control	Identity and access control	Signature Wrapping Attack	Data should be transmitted via a secured channel and fine-grained. Authentication and authorization techniques, hierarchical identity-based cryptography (HIBC)
Network Level Security Issues [64, 122]	Data confidentiality	Eavesdropping, Port Scanning, Replay attack, Sybil attack,	IPsec Implement security policies. Clearance of Old ARP addresses from the cache. Proper configuration, MD5/TTL protection Restriction of ICMP and SYN packets on the router interfaces. Domain name system security Extensions
	Data availability	Reused IP address, DDoS attack, BGP Prefix Hijacking, DNS Attacks,	
	Data Integrity	Sniffer Attack, DoS attack	

3.4.1 Fine-grained access control

Fine-grained access control commonly used method in the cloud environment. Using this mechanism every data entity is assigned its own access control policy. Every user can access only its data for which he is allowed and he should not be able to access data for which he is unauthorized. One who wants to access the data entity requires to provide its authorizations to a policy enforcer (not data owner in the cloud). For example, when an organization keeps data in the cloud, the organization allows only some authorized persons who are related to the projects, can see the data. The access control policies and the authorizations may disclose some information that is not allowed for the policy enforcer. To control the access to susceptible data or code fine-grained access controls must be accessible. In cloud computing, Cryptography is a good option to attain fine-grained access control [124]. In these schemes, Attribute-Based Encryption (ABE) is applied for data encryption. The decryption only can be performed by those who have the required attributes. These types of access control schemes are created by the storage service provider to store the data. To cloud resources control accessing, eXtensible Access Control Markup Language (XACML) standard is used for access policies. Some examples using XACML are GoogleMaps and salesforce, these providers use it for access control and authorization selection. The one challenge with the fine-grained access control methods is the policy enforcers can see partially the access control policies.

3.4.2 Privacy-preserving

Cloud computing is very powerful as compared to personal computing but the cloud also comes with new security issues to the data of the users. Data security and user privacy meet with various threats. Lots of work have been done to preserve the privacy of the user. The privacy information of users such as user credentials should not be revealed to the cloud. To understand the powerful and privacy-preserving service of data sharing in the cloud environment, some requirements must be attained such as the data possessor must be able o take decide that who can access his data in the cloud. Another is, the user's privacy must be secured in the cloud and lastly, the data must be accessed by low computing devices such as tablets and smartphones, etc. Dynamic accumulator-based technique for privacy-preserving access control has proposed by Slamanig [79], in which permissions read, delete, write are given by using Access Control List (ACL) to other users who can do tasks on outsourced data elements with mentioned permissions. With this technique granting or denying access can be decided. The user of the data can allow or gain access permissions to/from other users, although the Cloud service provider cannot recognize these users. The drawback of this technique is that if the owner of the data wishes to repeal permission from the user, then that user has to repeal given permissions from other users. This is complex computation to manage the chain of users.

3.4.3 Collision resistance

No user should not be allowed to share her/his private key user. Users cannot decrypt encrypted data by merging their features because each feature is associated with a random number or polynomial. Merging features from several sets of features within a specified key is an actual issue. Avoiding collision by ignoring users from merging features from several keys is another issue. Park et al. [125] presented Sec-DPoS a deduplicatable proof of storage system which is based on the symmetric key that guarantees confidentiality with brute-force attack resilience. It supports symmetric key cryptography-based integrity auditing of the outsourced data. They described some building blocks such as collision-resistant hash function, pseudorandom function, key derivation function, and pseudorandom permutation, authenticated encryption, deterministic symmetric encryption.

They have four protocols in their system. Key and index distribution protocol, initial upload protocol, and the Deduplication Protocol , and finally they discuss the integrity auditing protocol. In key and index distribution protocol, a preliminary uploader creates a possible response set for a file to audit Integrity with the help of a message-derived key that is distributed from the management server. The file and possible response set is uploaded. The other customer who is using the same file then he can use the possible response set created by the previous uploader. In the initial upload protocol, it is considered that the uploaded file is new data not earlier uploaded. Therefore, the customer creates the possible response set which will be used in the integrity auditing and another possible response set is created by the cloud server that will be used for ownership check. In the deduplication protocol, the deduplication procedure considers that the uploaded file is a duplicated data from an earlier upload. Therefore, the cloud server must check that the customer has the file. And in integrity auditing protocol, the customer who is the file owner then he can audit outsourced data's integrity at any time.

3.5 Privacy Issues in Cloud

Cloud computing uses virtualization technology, so, the personal information of the users may be dispersed in several virtualized data centers. When users access cloud services, they may disclose private information. The major goal of cloud security is securing data privacy. It is hard to avoid threats in a cloud environment due to its sharing environment that relies on a shared infrastructure. Therefore, data will be unprotected from unauthorized access. There are some privacy-related issues are discussed below.

3.5.1 Defining roles to actors

In the cloud environment, there are many types of the actor with different privilege. The main actor is the owner of the data, which can be called the data owner. Another important factor is the cloud service provider who can process private data like transfer, storage, research duplicate, etc and able to perform undeclared activities that can be a threat to data privacy like they can reveal private data by reading and also they could obtain user's data that can be useful for big financial advantage. Cloud-based services actors that may require to access private information to perform some actions and is the owner of the service who have the access to the database from where he can access the private information of the users [126].

Some other actors are cloud auditor, cloud agent, and cloud carrier. The cloud carrier is a mediator that makes available connectivity and transportation of cloud services from Cloud service providers to cloud Customers. The cloud broker is the individual who controls the routine, performance, and provision of cloud services and negotiates associations between service providers and cloud customers. The cloud auditor is the entity that can perform an independent evaluation of cloud services, performance, information system operations, and cloud security employment.

3.5.2 Compliance

Compliance states the responsibility of the company to work in agreement with existing standards, rules, and regulations. Different countries have their security and privacy rules and regulations that make compliance very complex and it becomes a critical issue in the cloud. Data location is a big challenge for the organization in compliance. The Service Level Agreement is important in the cloud base system. This is the agreement that is signed by the communicating parties that contain rule and regulation and all service information. There are various compliance issues, such as data sites or laws and regulations. Data site is the general compliance challenge encountered by the companies [127]. The organizations that use an internal computing center permit to design their computing environment.

Hence they have the detail information about the data storage and what security is being used, whereas, in various cloud computing services, data are warehoused in many physical sites, and complete information about the site of the data of the companies is not available or not open for the service users. This condition makes it problematic to determine whether adequate protections are in location and whether legal and governing compliance needs are met, such as NARA regulations that have ability requirements federal records storage and instruct the least height above and move away from a flood plain. The other issue is law and regulations, for U.S. Federal agencies, the key privacy and security compliance involves the Office of Management and Budget (OMB), Clinger-Cohen Act of 1996, 1974 privacy Act, 2002 E-Government Act like FISMA .

3.5.3 Legal issues and multi-location issues

The cloud business model adopts Service Level Agreement to specify the contracts over a particular service. This service might be SaaS, PaaS, or IaaS. The Service Level Agreement is contracted to legally decide the cost per service. Therefore, there can be an indirect subjectivity on the attainment of these contracts. The cloud actions increase several legal and challenges. The most important is the multi-location specific properties in cloud computing. Cloud service providers have the data centers in which they have sufficient resources to be spread worldwide. Irrespective of that, few countries do not permit data to leave their borders. Due to this situation, many challenges arise. If data moves beyond the boundaries, it is difficult to know due to which country jurisdiction data crashes. If it happens, it is difficult to declare to which area legal jurisdiction can reach to discover liable parties. It contains evaluating that government departments are able to access the data outside the boundaries of the country in which information was created in the first place.

3.5.4 Privacy issues on CIA

Privacy of the data has issues on integrity, authorized access, or availability. When information is copied or stolen by an unauthorized user , this condition is known as a confidentiality loss. When information is changed in an unexpected means, this condition is known as integrity loss. When information is missing or not accessible, this condition is known as availability loss. The data integrity provides the surety that data has not been modified during communication. Authorized access avoids data from attacks whereas backups and copies permit access of data properly even technical issues. Data is communicated at the common network backbone. Therefore, attackers can position concealed proxy applications between the Cloud service provider and customer to look for session detail and login credentials information of login. Attackers may do IP-spoofing or packet sniffing and they can get access to sensitive information.

Cloud services must be available all the time. In Infrastructure-as-a-service, the availability of hardware and logical resources such as computing servers and databases are required to run processing operations of programs and operations related to obtaining the data, respectively. Subashini et al. [101] discussed that a multi-tier design must be embraced that is reinforced by a load-balanced farm of application instances operating on various servers. This method lets DoS attacks resiliency by developing software and hardware breakdown measures in every tier.

3.5.5 Protection of the data

In the public cloud, the gathered data exists in the shared environment. Organizations keep sensitive data in the public cloud. There is a requirement of mechanisms in place so that the data access can be controlled and the data is stored securely. In today's era, data is currency and cloud storage is bank safe. Data is becoming gradually favorite target because of collective values [128]. The companies that have gathered data is at high risk and may cause DoS, as an accidental loss from an attack targeted against the companies. The physical attacks against the cloud resources of prestigious companies have side effects. For data security, data sanitization and isolation of data approaches can be used. Data can be found in various forms such as application development in the cloud and data can be scripts, programs, etc.

For installed applications, it contains records and other data produced or used by the applications. The access control methods are one way to keep data away from unauthorized clients. Furthermore, the method is encryption. Access control can be client's identity-based authentication that is a critical challenge in the cloud. Encryption is the only method to ensure the security of the data as less physical control over data storage. The data sanitization performs that a cloud service provider implements have apparent security effects. Sanitization contains erasing data from storage by overwriting or demagnetizing, or destroying media itself to avoid unauthorized exposure of data.

3.5.6 User control lacking

The private data are warehoused in the public cloud in distant machines that are controlled by the cloud service provider. There is a need for transparency for storage location, copies of data, and data processing. Sometimes it is almost impossible to get to know about privacy violations and who did them. In the cloud computing environment, computation and customer's sensitive data sharing do not have sufficient control, heading to threats such as stolen, exploit, or unauthorized access [129]. The software-as-a-service platform provides control of the data to the service provider, so the data control and visibility will be restricted. The attacker can steal or corrupt the data as users do not have control over the cloud. Moreover, there is no data transparency, for instance, data location, ownership of data, and data usage. Though data disclosure can occur during data moving, various countries have the law for data accessing if they suspect. A user control can be either a legal problem or one raised by the user himself.

3.5.7 Data movement

Data movement can occur between countries and local rules. Hence, it is another big challenge in the cloud environment. Data invisibility can be the solution for the privacy and security of the data of users. Fatema et al. [130] presented a . This model provides visibility into the place of data. XML-based policies are used in this model. The one drawback is a single point of failure. The central controller is inside the cloud and manages all data accesses and also data movement. If the accesses are made inside the same cloud, it can be the bottleneck. They consider inter-cloud data transfer, but every cloud needs to be managed own policy archive. The Directive of the European Union to secure the data (95/46/EC) related to moving personal data to the third countries came into existence in 1998 that permits moving sensitive data to third countries. The regulation and directives of the European Union are used to tie its member states. As an outcome, the United States Department of Commerce proposed a contract to permit the United States-based organization to share private data with their European complements without disruptions.

Some other privacy issues such as resource privacy sharing, sharing resource privacy in the cloud environment with protecting users is a significant issue. Data isolation can be used as a solution to this issue in which isolation is provided for each user's data, and one more big privacy issue is the data source nature as data comes from various sources. So, it is necessary to secure and control the data carefully, and only authorized users can access the data. Privacy of users must be controlled when data is gathered, warehoused, or carried. Data gathering from various sources is challenging in cloud computing because it exposes sensitive information of the users. To attain an adequate level of privacy in cloud, various problems need to be discussed, such as inadequate customer control over their data, data revealing in-transit across the cloud, illegal secondary storage of delicate data, uncontrolled data propagation, and dynamic provision of legal difficulties [131]. One other issue is the advanced service level agreement method relies on the customers' feedback concerning significant challenges such as Cloud interoperability, data, resources, and processes that can be moved to the cloud.

3.6 Conclusion

Cloud computing has evolved over the years. Security is one of the primary concerns in cloud environment. In this chapter, various definitions of security and privacy are provided first. Furthermore, various cloud security goals are discussed. Various issues related to cloud security are explained in detail with viable solutions. In addition, security requirements for the privacy are explained to give reader deep understanding on privacy aspects in cloud. Furthermore, an various privacy issues are discussed. We hope that this chapter will give a good knowledge about the security and privacy concepts in cloud.

3.7 Questions

Fill in the blanks

1. Mark the correct privacy issues in cloud
   i Protection of data
   ii Lacking of user control
   iii Data movement across countries
   iv All of above
2. Data linage refers to trace the path of the data and this is essential for auditing in the cloud.
   i Trace the path of the data and this is essential for auditing in the cloud.
   ii Unauthorized use of data in a shared environment
   iii Unwanted modifications in the data
   iv Unrestricted access to data
3. The unavailability of services is the
   i Network level security issue
   ii Application level security issue
   iii Data security level issue
   iv Virtualization level security issue
4. Access control features can be used
   i To maintain the control on access to the resources.
   ii To know what is going on in cloud
   iii To enable the remote data access
   iv None of the above

5. Mark the incorrect statement

   i Data integrity is the basic task that verifies the data and it provides the guarantee for the exactness and quality of the data.
   ii The goal of availability is to provide the services to its users without modification.
   iii Authentication is the method of creating assurance in the identities of the user.
   iv Auditing is the monitoring task to know what is going on in the cloud-based system.

Short-Answer Questions

1. Describe application-level security issues.
2. Why privacy is a research challenge in cloud? What are the security requirements of privacy.
3. What is the difference between confidentiality, integrity and availability?

Long-Answer Questions

1. Define privacy. What are key privacy issues in cloud?
2. Describe virtualization-level security issues and data security issues in cloud.

Part II

Threat Model, Attacks, Defense Systems, and Security Techniques

Chapter 4

Threat Model and Cloud Attacks

4.1 Introduction

Cloud security has become the area of global importance and it is being widely discussed across the academia and industry. On one hand, the sharing and distribution of cloud resources make it easier for the attacker to target the cloud component. On other hand, such a multi-tenancy nature of cloud makes it difficult for the developers to design a security model to ensure the security and privacy of services. Some of the VM-level attacks may be handled by the tenant's security tool. However, there are many advanced attacks which can even breach the VM security tools. The advanced attacks can bypass the guest OS kernel security and try to access the VMM-layer illegitimately. The CSPs can provide the security from attacks by extending the security into the virtualization-layer. There may be chances that security breaching incidents, threats and attacking reports are not shared with customers. Hence, CSPes usually highlight the security and privacy concerns of cloud service in service level agreement (SLA) [64].

The existing threats and attacks in cloud environment may harm the cloud services in a direct or indirect way. The availability, integrity and confidentiality of cloud resources is affected which raises a very serious security concern. Hence, it is the responsibility of the CSP to secure various layers of cloud environment. Many researchers are working in this direction and designing various security methodologies that can be adopted by CSPes.

However, it is essential to first identify various threats, attacking entities, attacking scenarios and various attacks before designing any security solution. This chapter aims to provide a detailed description for same. We have considered the OpenStack-based architecture [7] as a base model for cloud environment for explaining various threats and attacks. OpenStack is a collection of open source cloud components used for developing cloud platform for public, private and hybrid cloud [132]. Initially OpenStack in jointly launched by Rackspace Hosting and NASA, as an open-source cloud-software in 2010 [133]. Later, SUSE, Red Hat, Oracle, IBM and some other companies launched their OpenStack-based distributions. VEXXHOST [134] provides public cloud services, powered by OpenStack.

AURO public cloud [135] offers IaaS and is also powered by OpenStack. RackSpace public cloud [136] also uses OpenStack for implementing the infrastructure for cloud services. DataCentred public cloud [137] is another solution for implementing IaaS services which is based on OpenStack. ELASTX OpenStack:IaaS [138], Dualtec [139], AgileCLOUD [140] are other examples of OpenStack-based public cloud. Rackspace private cloud [136] and IBM Bluemix private cloud [141] are

DOI: 10.1201/9781003004486-4

also powered by OpenStack. In the OpenStack summit 2016 at Barcelona, Platform9 [142] announced its contribution toward hybrid cloud implementation, using OpenStack. They extended the project and included OpenStack drivers for Amazon Web Services (AWS) to integrate the OpenStack services (nova-compute, neutron, etc.) with AWS.

The openstack components are as follows: Cloud Controller Server (CConS), Cloud Compute Server (CComS), and Cloud Network Server (CNetS). There are mainly three types of networks in cloud environment: management network, tenant network, and external network. All the private data of tenants travel through the tenant network. There are various tenant networks created by the providers, which are in charge of transporting the data of a set of tenant users. Each tenant network has a unique id. The tenant machines are connected to the outside world through external network (internet). Tenants can connect to the cloud server from any location via external network. The more details about the architectural components and the role of various members are described in detail in Chapter 1.

4.2 Threat Model

Threat model is used to illustrate and understand the assets, which can easily be targeted in cloud environment. It also helps in understanding various attacking surfaces in cloud that can be exploited by attackers. In this section, various attacking entities along with attack surfaces and attacking scenarios have been discussed in detail.

4.2.1 Type of attack entities

To take the benefits of cloud computing, users/enterprises have to move their application into cloud. Once on premise applications are migrated into cloud, the users lost their physical control on data. The application is now deployed in open computing environment and can be exposed to various attacks. It is crucial to know about attackers and how they can attack so that prior precautions can be taken against them. Let us first understand the meaning of insider and outsider terms. Insiders are the organization's people who could be administrator, owners, employees and staffs, etc. Insiders may also have the physical access to resources (directly) or may have contacts with organization's other employees having physical access (indirectly) are people who are associated with the CSP's organization from outside. For example, people who have registered to cloud services and using/not using the cloud services and people who have not registered to CSP's organization. In addition, the third party resource provider to the cloud organization if some resources are outsourced by CSP and brokers, etc., are also considered as outsiders. There can be many ways to classify the attackers [143]. Here, the categorization is done based on the privileges of cloud resources; i.e., (i) insiders having highest privileges, (ii) insiders having moderate privileges, (iii) opportunistic insiders, (iv) outsiders with limited privileges, (v) outsiders with no privileges. The discussion on each of the category is given on the following pages.

A. Insiders having highest privileges

Cloud administrators and cloud service providers are the biggest threat to the data stored in cloud servers, if they are not trustworthy. They have the highest privileges to all the cloud resources. In fact, cloud administrator has got all the credential information of the cloud users. If such insiders become malicious insiders, they can intentionally misuse such privileges in accessing any of the virtual machine running in cloud, leaking or modifying the user's data stored in cloud. Due to lack of complete transparency, the administrators can also show a false good configuration to virtual machine users which can lead to higher computing or resource-usage charges. The user's credentials can also be shared with unauthorized users to gain access the sensitive user's information. The malicious insider can also try to hind the data which is corrupted by some server failure and hacks. He/she can also deliberately delete the rarely used data of the customers just to free some resources without taking the concern from user.

B. Insiders having the moderate privileges

Cloud service provider's employees who have been assigned some roles such as Cloud developers, cloud system engineers, cloud network engineers, cloud product manager and other technical employees of cloud team, have been some privileged (moderate) to access the cloud resources. A malicious cloud employee/staff can be threat to a company. A malicious employee can leak the customer's data under his/her access to third party resources. The highest level of abstraction is provided to outsiders from the cloud activities. As a result, malicious employee can take the benefit of this feature and can hide the effect of some malicious event executed by them. A malicious employee can even take control on customer's virtual machine and try to collect the information about virtual machine user. A network administrator can monitor the communication logs of virtual machines and extract some useful information.

C. Outsiders with limited privileges

Cloud customers have access to the virtual machines or services which have been subscribed by them. Once the customer gains access to the cloud network, he/she can try to collect certain information of other users using the limited privileges allocated to him/her. There are some attacks such as sniffing attack, scanning attack, etc. which can be launched by malicious customers. A malicious cloud customer can sniff or scan the cloud network traffic which is passing through virtual switch which connects the VM to other co-residing VMs to extract some useful information about VM. The information then can be used to launch further attacks. In addition, a malicious cloud customer tries to bypass the limited access given to him by running malicious codes inside the VM allocated to him/her. The attacker can thus gain root access to the hypervisor kernel which can lead to many destructive results. Once a hypervisor is compromised, it can become a serious threat to all the VMs, running in the same server. VM Escape is one such malware attack in which attacker bypass the memory access allocated to it and tries to access the host memory or hypervisor memory or other VM's memory.

D. Outsiders with no privileges

Most of the cloud services are deployed in public domain which can be accessed through internet. It is vulnerable to various potential attacks from internet. Most of the cloud service providers only require some basic information including credit card information from users to register them with cloud services. There is no way of finding the intention of a user while creating a user account in cloud. The attacks launched by registered users who have some privileges have been discussed above. However, outsiders having no cloud account can also try to launch cloud attacks. For example, attacker can create phishing website of cloud service provider in order to collect the credit card details of legitimate users and cause monetary harm to users. The network traffic exchanged between legitimate users and cloud servers can be eavesdropped and manipulated by outside hackers. The cloud service provider's website can be prone to various web attacks (sql injection attack and cross site scripting attack (XSS), etc.), which raises a strong concern for security. Once the administrative server is breached, the credentials of legitimate users can easily be hijacked by the outside attacker.

The description about various attack surfaces along with attacking scenarios is given in threat model, described next.

4.2.2 Attack surfaces with attack scenarios

In the threat model, a circle symbol is used to represent the start of the source and arrow is used to represents the destination/victim machine. The 'from source-to target' representation depicts various attacking surfaces and are shown using numbered lines. The threat model has been proposed by us in previous work [144], shown in Figure 4.1. Let us explain each attack scenario below.

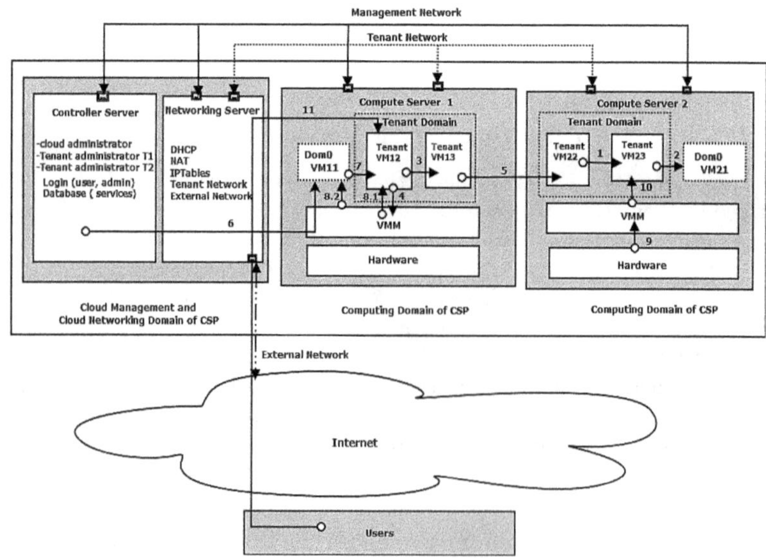

FIGURE 4.1: Threat model in cloud environment.

Scenario 1

A Tenant Virtual Machine (TVM)'s user can try to access another TVM maliciously by using the privilege escalation techniques. Once an attacker becomes successful in bypassing the access of another TVM, he/she can execute the rootkit malwares in guest machine in order to gain the root access of the victim machine. A rootkit can hide the intrusions and executes with higher privileges of guest OS. Such malware can cause the harm to the victim VM. Some of rootkits are evasive in nature and can subvert the security analyzer running inside the victim VM. These types of attacks are called as VM-VM attack (shown in line no. 1).

Scenario 2

Some of the malware first try to bypass the root access of the allocated VM and then run the advanced malicious code to cross the memory boundaries beyond the access of the VM. The attacker can be successful in gaining the root access of the privileged domain (i.e., Dom0) of VMM by executing such malwares. Dom0 is the administrative VM and used to perform all management commands for defining the security policies for the VMs running above it. Any compromise at the hypervisor-level can breach the security of all the VMs running above it. VM Escape [91] is one such attack.

Scenario 3

TVMs can communicate with each other. For example, VM13 can communicate with VM22 if provider has provided them free communication channel. A malicious tenant can exploit the channel and use it for performing malicious activities such as scanning and flooding. DoS attack is one such example where victim's server is flooded with too many network connection requests. The network resource starvation at the victim machine may become cause of DoS to other co-located VMs . The service denial can also lead to the violation in services level agreement (SLA) (shown in line no. 3).

Scenario 4

Tenant VMs of different tenant administrator can be allocated on same physical server. The co-residency can be exploited by the attacker and co-located VM can become the victim of attack. Let us consider: VM12 belongs to a tenant administrator (TA1) and VM13 belongs to tenant administrator (TA2). As VM12 and VM13 reside on same server, a malicious tenant user can misuse the access to VM12 and perform scanning activities to fetch the details of other co-located VMs such as VM13. This attack is called as VM to VM attack (shown in line no. 4).

Scenario 5

A malicious user can compromise the guest OS root privileges of the VM allocated to him/her. He can further run some malicious programs to gain access to unauthorized VM memory regions. A VM can also cause the resource starvation at the virtualization-layer. Once all the resources at the hypervisor-layer are exhausted, hypervisor will deny providing resources to other VMs running above it. It will lead to denial of service (DoS) at the VMM-layer. This attack is called as VM to VMM attack (shown in line no. 5).

Scenario 6

Infact, network attacks can be launched in cloud in a more complicated way. A malicious tenant can generate IP/MAC spoofed network traffic of the victim virtual machine and flood such packets in the network. All the VMs will now reply to the victim machine causing the network resource congestion at the victim server. It will cause denial of service at the network level and may lead to clashes between CSP and victim user (shown in line no. 3 and line no. 4).

Scenario 7

Attacks can be launched by malicious CSP, called as insider attacks. A malicious CSP can misuse the privileges of accessing the privilege domain (Dom0) of VMM of the VM hosting servers. He/she can access the root terminal of privilege domain and try to access the VMs running above it. Once VMs are accessed, unwanted malicious scripts can be executed at the guest OS to perform some malicious activity with an intention to harm VM. Such attacks are called as VMM to VM attack (shown in line no 6 and line no 7). However, generally service providers are much concerned about their reputations and hence will avoid such attacks which can destroy their relationships with tenants. A feasible solution to select a trustworthy CSP has been explained by Habib and Varadharajan [78].

Scenario 8

Some of the attacks can be targeted toward cloud infrastructure by users through the internet. A malicious outsider may try to gain access to the cloud resources by registering himself/herself with CSP. Infact, an outsider can also try to gain illegitimate access to a VM running in cloud, causing the unnecessary utilization of VM resources. Victim suffers by paying for the additional charges incurred which causes dispute between service provider and tenant user. An attacker can also exploit the billing model of service provider which will generate the incorrect bills for the tenant VMs by making the victim's resources busy. It is called 'economic denial of service (EDoS)' attack. A tenant has to suffer for paying the additional billing charges ((shown in line no. 8).

Scenario 9

In addition, CNS can also become the victim of attack as it is the main point of contact for all the traffic passing through Cloud network. All the inbound and outbound network traffic passes through CNS. It can be subjected to flooding, scanning and brute force attacks which will degrade the performance of Cloud services. Once

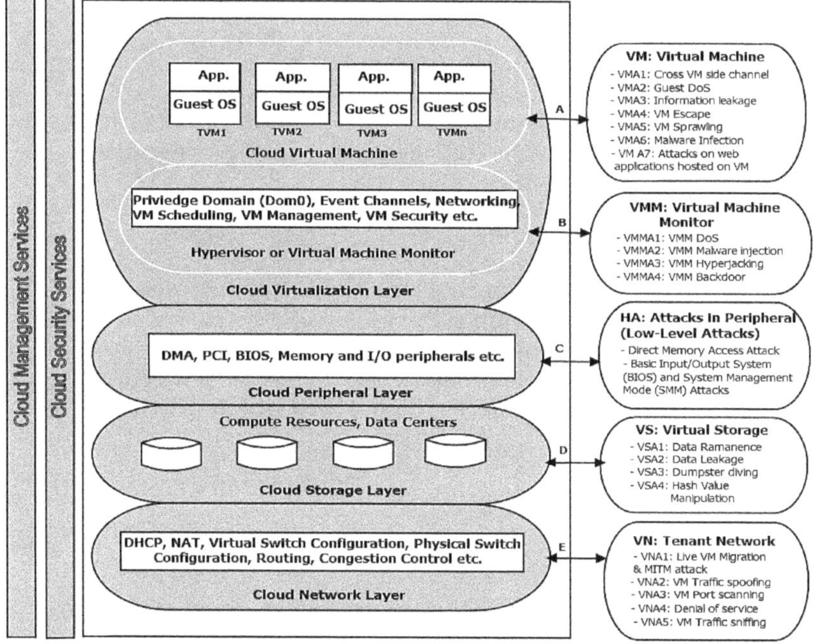

FIGURE 4.2: Attack taxonomy in a cloud environment.

CNS goes down, none of the cloud services can be provisioned to the customers (line no. 9).

4.3 A Taxonomy of Attacks

The attacks can be classified based on node-level, layer-level and/or target. Here, the classification is done based on the target component as per our exhaustive review done in our previous work [144] (refer Figure 4.2). Attacks specific to virtual environment have been considered. A brief description about various attacks has been described in this section next.

4.3.1 VMAT: Virtual machines-level attacks

Virtual machines (VMs) are one of the most critical cloud resources that could be easily bypassed by attacker (line no. A) in cloud because of its easy accessibility. Some of the VM-level attacks are explained below:

VMAT1: Cross-VM Side Channel

In this attack, the side channel information such as time, cache, power and heat, etc. are extracted and the private key information of victim VM stealed. This attack breaches the isolation between two VMs. As an example, CPU cache is one the shared resource which can be breached. Some of the cache-based attacks Trace-driven, time-driven and access-driven [58]

VMAT2: Guest DoS

Any malfunctioned Hypervisor can serve all the malicious or non-malicious request of VMs for the allocation of resources. If a VM is allocated with all available resources, it will become impossible for the hypervisor to allocate any more resources. It will cause starvation of resources to other VMs. It leads to slow down the process of some of the critical applications. For example, VMWare Workstation 6.0 can be prone to DoS attack though a device driver, called [145].

VMAT3: Information Leakage

There are different ways to leak the information from a VM memory. Let us consider a case in which VM information can be leaked using the malicious use of introspection functionality at VMM. A malicious cloud administrator can gain access to the VM memory from VMM using the introspection libraries. In fact, a malicious user who has compromised the VMM can misuse the root access to VMM in gaining illegitimate access to VM memory.

VMAT4: VM Escape

In this attack, an attacker bypasses the memory which is allocated to its VM, and tries to reach to the illegitimate address regions in memory beyond its access (other VMs, Host OS or VMM memory). This attack can also be considered breaking out of VM and talking to underline host or VMM. In addition, the isolation of VMs cab also be breached by attacker machine which will lead to discloser of the private data of VMs, leading disputes between service provider and service consumer [91].

VMAT5: Malware Code Infection

A program which behaves maliciously is called a malware that cab be executed as a self-explanatory code or injected with some legitimate program. Some examples of such malicious code are trojan, virus, rootkits and worm, etc. These malware codes can perform some harmful operation leading to the threat of integrity of the system. Some of the malware can even replicate themselves and can crash the OS. Malware can be injected in the guest OS to gain root access to the target VM. Once a root access to a VM is gained; further attacks can be launched.

VMAT6: Web Attacks

The enterprise cloud environment can easily be subjected to a variety of web-based attacks like phishing attack, cookie modification attack, etc. An attacker can attempt to access confidential data of customers by hosting a malicious phishing website on the cloud servers. These malicious websites can be used to gather the sensitive data of customers. Raytheon has become the victim of such cloud-based phishing attack [146]. The vulnerabilities in web services can also be exploited by attackers such as vulnerability in SSH.

4.3.2 VMMAT: Virtual machine monitor-level attacks

Attacker can also exploit the vulnerability present in the hypervisor code (line no B) in taking control of VMM kernel. Once a VMM is compromised, it can perform harmful operations and gain access to the VM memory. Some of the attacks at the VMM-level are described below:

VMMAT1: VMM DoS

Resource starvation at the VMM-level such as RAM, CPU, network BW consumption can badly affect the operations of VMM. VMM operations can also be interrupted by restarting VMM for each new VM service. Both the situations can lead to DoS, affecting the VMs running on top of it [147].

VMMAT2: VMM Malware Injection

A virtual environment can even be sensed by a malware. Once a malware came to know about it, it tries to disable the critical analysis component such as VMM. VM can be abruptly rolled back by malware and can be restored to previous point.

VMMAT3: VMM Hyperjacking

In VMM hyperjacking attack, a rough hypervisor is inserted above the real hypervisor layer which takes complete control of the system. A VMM-based rootkits malware are usually used for launching such attacks. The features of virtualization such as AMD-V and Intel-VT X are exploited by rootkits to gain higher privileges of the system. Some of the examples of such rootkits are Subvert and BluePill [69].

VMMAT4: VMM Backdoor

A backdoor entry can be taken by an attacker to escalate the privileges of the hypervisor either by altering the kernel level data structure or overwriting some sections of VMM code. Such attacks are very difficult to detect if they are launched from the privilege domain of hypervisor. There are two backdoors which have been implemented [148]: first through VMM code and other through hidden domain with compromised privileges.

4.3.3 HWAT: Peripheral–level attacks

Once an attacker has physical access to a memory, he/she can launch hardware threats. Some of such threats which target the integrity of the tenant's data and are launched at the peripheral-level (line no. C) are explained as follows:

HWAT1: Direct Memory Access (DMA) Attack

In order to provide faster access to memory, Direct Memory Access (DMA) concepts have been introduced some years ago. However, DMA can be subjected to malware attacks by using malware such as DMA malware. These malware attacks run in the hardware and try to bypass the host security. These types of attacks can try to gain access to information such as cryptographic keys in order to access user's confidential information from secondary device. For example, one of the DMA-based malware DAGGER [149] targets Windows and Linux.

HWAT2: System Management Mode and Basic Input/Output System-based Attacks

The system security and power management functions are managed by System Management Mode (SMM). It is one of the highly privileged modes of CPU. This mode can be prone to attacks such as cache poisoning attack Which tries to inject the malware code at System Management RAM (SMRAM). The system cache lines of SMM code can be attacked [150]. Any vulnerability present in BIOS can compromise the SMM code and make it possible for an attacker to launch further attacks and breach the isolation of VMs.

4.3.4 VSWAT: Virtual storage-level attacks

The sharing of physical storage among many tenants improve the system utilization features. However, such a multi-tenancy can be exploited and following attacks can be launched (line no. D).

VSWAT1: Data Remanence

The remaining information about data which prevails in storage even after deletion corresponds to Data Remanence. It can result in leakage of user's sensitive information. If user removes the data from the virtual storage then CSP should ensure data remanence should not occur [59].

VSWAT2: Data Leakage

Data leakage can be performed by different ways. There are various tools which can be used to gain the access of credential of users to access the target VM such as key logger. Data can also be leaked by malicious service provides to some third parties, leading to disputes. VM Backdoors such as VMware I/O backdoor can gain access to VM data [151].

VSWAT3: Dumpster Diving

Attacker attempts to extract the useful information from the deleted data such as trash, recycle bin and try to recover the user's information from it. The data which is marked as waste by cloud admin or user can provide useful information such as credit card details, cookie information, key details, etc. [59].

VSWAT4: Hash Value Manipulation

The hash values of the files which are stored in cloud servers can be manipulated by attacker. The hash value now contains the modified hash value which now points to the malicious file. The vulnerability can also be found if server operates on OpenSSL by making use of wrapper classes like Ncrypto [152].

4.3.5 TENAT: Tenant network-level attacks

Network attacks are also possible in the cloud environment which targets the network vulnerabilities (line no. E). Some of the examples are given below.

TENAT1: MITM Attack and Live VM Migration

A malicious user can also try to evade the network communication between two VMs. There are higher chances of data eavesdropping when VM is in migration stage. The data leakage during transit stage is also called as Man in the Middle (MITM) attack. Attack breaches the private key of victim machine and forms the communication channel with sender [153].

TENAT2: VM Traffic Spoofing

The VMs can also become victim of spoofing network attacks. IP spoofing is one of the famous network attacks in which attacker's machine generates IP spoofed network traffic and floods it in network with the source address of victim VM. The rest VMs in the network then start sending replies to the victim machine, causing resource exhaustion [70].

TENAT3: VM Port Scanning

Port scanning attack is performed to scan and obtain the useful information of target VMs like list of open ports, version of OS, version and name of software installed, etc. inside the target machine. It helps in launching further attacks. Some scanning tools are hping3 [154] and nmap [155].

TENAT4: Denial of Service

Attacker performs the traffic spoofing or simply flooding at the broadcast network. The flooded traffic contain the source IP address as the target victim VM's IP. After receiving such traffic, each server in network responds to victim VM, causing the consumption of resources and hence denying providing the cloud services. If more than one machine carry out for flooding to cause DoS, it is called as Distributed DoS (DDoS) [156].

TENAT5: VM Traffic Sniffing

VM traffic sniffing can be carried out at the open virtual switch (OVS) level. VMs share the same hardware resources of physical machine and are connected through virtual bridge and virtual switch. Attacker can performs the sniffing easily from the compromised VM to know more details about other VMs running in same network.

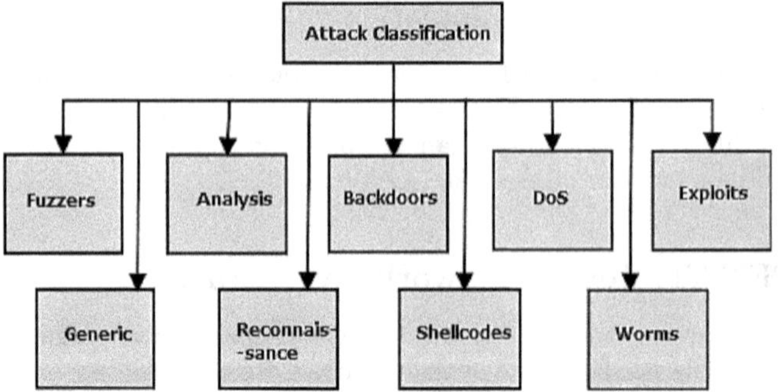

FIGURE 4.3: A taxonomy of various attacks based on UNSW-NB dataset.

4.4 Case Study: Description of Features for Attack Analysis Based on Dataset

Most of the researchers, have considered attack dataset such as KDD99 [157] and UNSW-NB attack datasets [158] for validating their approach in cloud. In this Section, we have provided a case study on attacks based on UNSW-NB dataset as shown in Figure 4.3. The important features for each attack are described under each category based on UNSW-NB dataset [158]. Each of the feature is described in detail in Table 4.1. As per UNSW-NB attack dataset, the attacks are categorized into nine categories (refer Figure 4.3). We earlier discussed about the denial of service attacks in details. Remaining attacks are outline beneath:

4.4.1 Fuzzers

In this type of attack, there is a large input which is supplied from command line terminal. In this type of attack, loopholes or vulnerabilities in operating system, network or program are investigated. Resource may be crashed or may not provide services on time if the identified vulnerabilities are investigated.

Features for Attack Analysis

If huge numbers of packets are sent by the source using similar service protocol or at the same sink node's port number for a period of time, it may be the implication of fuzzer attack. Indeed, there are various features like source to sink bytes, number of packets from source of sink and variations in the frames (Jitter) may be the implication of problem. Some features are important in this case like dur, service, sbytes, spkts, srcjitter, synack, cf_srv_src, ct_src_dport_ltm, described in Table 4.1.

TABLE 4.1: TCP connection features of UNSW-NB dataset

Flow Features	
srcip	IP address of Source
sport	Port address of Source
dstip	IP address of Destination
dsport	Port number of Destination
proto	Transaction protocol
Basic Features	
state	The state and its dependent protocol, e.g., ACC, CLO
dur	Stores total duration
sbytes	Src to dst bytes
dbytes	Dst to src bytes
sttl	Src to dst time to live
dttl	Dst to src time to live
sloss	Src packets retransmitted or dropped
dloss	Dst packets retransmitted or dropped
service	ftp, http, ssh, dns
sload	Src bits per second
dload	Dst bits per second
spkts	Src to dst packet count
dpkts	Dst to src packet count
Time Features	
sjit	Source jitter (mSec)
djit	Destination jitter (mSec)
stime	record start time
ltime	record last time
sintpkt	Source inter-packet arrival time (mSec)
dintpkt	Destination inter-packet arrival time (mSec)
tcprtt	The sum of synack and ackdat of the TCP.
synack	The time between the SYN and the SYN ACK packets of the TCP.
ackdat	The time between the SYN ACK and the ACK packets of the TCP
Additional Features	
is_sm_ips_ports	If source (1) equals to destination (3)IP addresses and port numbers (2)(4) are equal, this variable takes value 1 else 0
ct_state_ttl	No. for each state (6) according to specific range of values for source/destination time to live (10) (11).
ct_flw_http_mthd	No. of flows that has methods such as Get and Post in http service
is_ftp_login ct_ftp_cmd	If the ftp session is accessed by user and password then 1 else 0. No of flows that has a command in ftp session.
ct_srv_src	No. of connections that contain the same service (14) and source address (1) in 100 connections according to the last time (26).
ct_srv_dst	No. of connections that contain the same service (14) and destination address (3) in 100 connections according to the last time (26).
ct_dst_ltm	No. of connections of the same destination address (3) in 100 connections according to the last time (26).
ct_src_ ltm	No. of connections of the same source address (1) in 100 connections according to the last time (26).
ct_src_dport_ltm	No of connections of the same source address (1) and the destination port (4) in 100 connections according to the last time (26).
ct_dst_sport_ltm	No of connections of the same destination address (3) and the source port (2) in 100 connections according to the last time (26).
ct_dst_src_ltm	No of connections of the same source (1) and the destination (3) address in 100 connections according to the last time (26).

4.4.2 Analysis

There are various intrusive attacks that can penetrate the web applications such as malicious web scripting, port scanning and transmitting spam emails, etc.

Features for Attack Analysis

We discussed several techniques such as port scanning, and detecting the threats in section III(B). To drain emails coming from unauthorized source, mail service providers, provide anti-spam filters. Some of the emails may bypass such kind of filters. Especially, by carrying out email header analysis, HTML header or email code analysis; some of such attacks can be recognized [159]. Thus, the investigation of overall network execution along with source IP address can be done by analysis of different probable features as listed in Table 4.1.

4.4.3 Backdoor

The normal authentication can be bypassed by gaining unauthorized remote access to a system using backdoor malware. To bypass the system security of the system, attacker efforts to locate the data by performing redolent activities. By putting malicious contents in files, transforming the code or gaining access to the remote system or data, some backdoor programs can be used by the hackers.

Features for Attack Analysis

Discrete some of the important features for such types of attacks analysis are: {sport, dsport, dur, sbytes, service, ackdat, sjit, djit, ct_flw_http_mthd, is_ftp_login, ct_srv_src, ct_dst_ltm}. Although it is very difficult to find out the information regarding a backdoor attempt at the victim machine. However, we can find some clue regarding the unauthorized network with the help of network analysis.

4.4.4 Exploits

There are some intrusion (an exploit category) that can affect vulnerabilities of software, bugs or glitch within the operating system or any other software. Attackers use various open attacking tools to launch payloads and exploit the specific vulnerability.

Features for Attack Analysis

There are several features those play a important role to identify or detecting the number of attempts of triggered exploits at monitored machine such as {ct_ftp_cmd, res_bdy_len, ct_src_ltm, ct_src_ltm } (refer Table 4.1). Some clue regarding the attempt of triggered exploits can be provided by these features. There are various other approaches which are used to detect such attacks within a machine using software or program behavior analysis [160, 161].

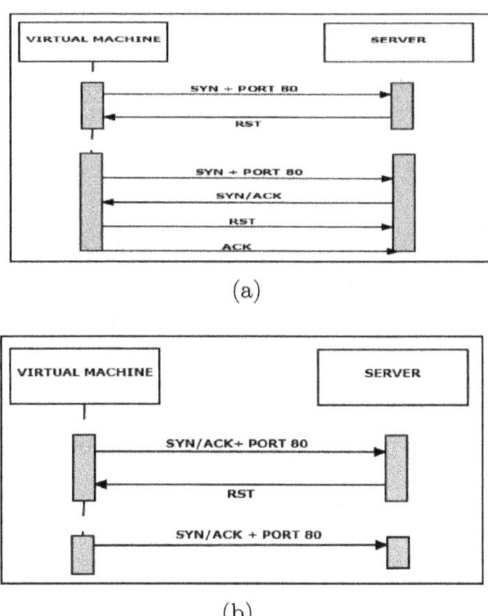

FIGURE 4.4: Open scanning to know about open close ports.

4.4.5 Generic

The secured key of the analysis system can be breached by generic attacks which try to evade the cryptographic algorithm. The generic attack do not require to have the technical details of the cryptographic system. For example, birthday attacks.

Features for Attack Analysis

For the consideration of generic attack, it would require all probable network features. But, if we only consider network features then accuracy of the system may be affected. The program code behavior can be monitored using advanced approaches. Some specific features are not provided by UNSW-NB like root_login, su_attempted, Hot, Num_Shell, etc.

4.4.6 Reconnaissance

The security control can be bypassed by collecting as much information as possible about the target network or victim device. This may be first step toward targeting the further attacks. There are various ways which can be used to gather the information, such as whois, dig, nslookup, port scanning tools, network scanning tools, etc. Once sufficient information is collected, attacker decides which attack to launch further. Figure 4.4 demonstrates the open scanning attack from a VM to Server. A series of messages with a specific flag value, are exchanged between VM and Server to know about open & close ports.

Features for Attack Analysis

There are several implications to identify such type of attacks: {sport, dsport, srcip, dstip, dur, spkts, sinpkt, service, synack, ct_srv_src, ct_src_ltm, ct_dst_ltm }.

4.4.7 Shellcode

To exploit the vulnerability of the software, a shellcode can be used as payload that runs at the target machine. A shellcode starts a command shell that can be control by the attacker. The highly privileged processes can also be exploited by shell codes such as buffer-overflow. The remote processes can be vulnerable to remote shell codes. If the remote shellcode run successfully, it makes the attacker successful in gaining access to the target system. For example, there is a bindshell which helps an attacker in connecting to the specific port of victim machine.

Features for Attack Analysis

There are several important features like:{sport, dsport, srcip, dstip, dur, service, sbytes, dbytes, state, res_bdy_len, synack, is_ftp_login} (refer Table 4.1). To detect the remote shellcode, some network features would helpful.To provide good accuracy and false alarm, shellcode can be detected with the help of program analysis techniques.

4.4.8 Worms

Worms are self-replicating malwares or malicious programs that duplicate itself to spread over computer to computer with the help of Internet. Although maximum worms cannot the system files and can replicates itself only. But they can try to distortion of the services by increase traffic.

Features for Attack Analysis

Some important features of the network are as follows: srcip, dstip, sport, dsport, proto, spkts, dpkts, tcprtt, stcpb, dtcpb ct_srv_src, ct_flw_http_mthd, is_ftp_login, etc. (refer Table 4.1) that would be helpful to analyse the packets.

A deliberate attempt to leak or modify the user's data or to gain unauthorized access to a machine is known as attacks. Some attacks can harm the system or network resources. Several attacks are discussed in this chapter. Each and every attack can be triggered by its own distinguish method and have some unique characteristics which we have already discussed.

The importance of each of the network feature is identified and set of features are mapped to specific category. In most of the research work, KDD'99 dataset has been widely used for validating the research methodologies. Here, we have considered a comparatively new dataset UNSW-NB [158] for the attack study.

4.5 Conclusion

Security and privacy in the new emerging era of technologies, is very important. It is very important to know about the architecture along with all design components and entities associates with cloud computing before doing any security analysis. We have thoroughly explained the basic cloud computing architecture to give readers an idea about the design components and members who are involved during an transaction in cloud. Openstack has been considered as a base model for explanation as many cloud service providers are following the openstack-based cloud computing architecture. A threat model has been explained to provide readers an idea about which all assets/components are likely to be targeted in cloud environment. Various types of attackers, attack surfaces with attacking scenarios have been described in detail. A detailed taxonomy of various types of attacks that can target cloud components have been provided. A case study on UNSW-NB dataset have been considered for discussing various attacks and associated attack features which should be considered while designing security techniques for detecting specific attacks.

4.6 Questions

Fill in the blanks

1. Web applications are prone to:

   i Cross site scripting attack
   ii SQL Injection attack only
   iii Cross-site request forgery
   iv All of above

2. Which of the following statement is incorrect:

   i Cloud service provider allow the cloud customer to impose their security model
   ii The live migration of customer data may be exposed to various network-layer vulnerabilities.
   iii Customers can install some security tools inside their own VM.
   iv Privacy can be defined as the obligation and right of an organization or individual

3. Distributed Denial of Service issues pertain to

   i Network level security issues
   ii Application level security issues
   iii Data security level issues

4. Data leakage during transit stage is called as

   i IP Spoofing attack
   ii Port scanning attack
   iii Man in the Middle attack
   iv None of the above

5. Mark the incorrect statement

   i Once applications are migrated to cloud, the users lost control on data
   ii Tenant virtual machine cannot communicate with each other
   iii A VM can also cause the resource starvation at virtualization-layer.
   iv Attack launched by malicious CSP is called insider attack.

Short-Answer Questions

1. Describe OpenStack architecture with access roles of different members.
2. What are Dumpster Diving, Data Leakage, VMM hyperjacking attack and VM Escape attacks?
3. Draw Attack Taxonomy. Differentiate between malicious insider and malicious outsider attack.

Long-Answer Questions

1. Explain features of USNW-NB dataset with their relevance for attack detection?
2. Explain the threat model and attack surfaces in cloud environment with suitable diagram.

Chapter 5

Classification of Intrusion Detection Systems in Cloud

5.1 Introduction

There exist various intrusion detection systems (IDSes) having different security architecture and working mechanisms in cloud environment. These IDSes can be classified into four major categories [64]: (i) Tenant Virtual Machine (TVM)-based IDS, (ii) Virtual Machine Monitor (VMM)-based IDS, (iii) Network-based IDS, (iv) Distributed IDS, as shown in Table 5.1.

TVM monitoring-based IDS performs monitoring at TVM-layer. At TVM-layer, the IDS is configured and executed inside a TVM. The visibility of monitored TVM is high with such IDSes. The audit analysis of host logs, program analysis of execution logs and static features (source code, byte code, opcodes, etc.) analysis are certain malware analysis approaches of TVM-based IDSes. On detection of any suspicious activities, these IDSes can be configured to send alerts to tenants. The key advantage of such IDSes is that the required information of monitored VM can be easily accessed by them. Therefore, TVM-based IDSes perform better when compared to other IDSes in terms of timely prediction and accuracies. No modification is required in the VMM code. However, they are less robust and have less attack resistant capability. They can be completely controlled and configured by tenants. As these IDSes are simple in design, they are suitable to be deployed in different cloud deployment scenarios. Bag of System Calls (BoS)-IDS [162] and Secure In-VM Monitoring (SIM) [163] are some examples of TVM-based IDS tools.

Hypervisor-based IDS performs monitoring at VMM-layer. Security tool is installed at VMM or priviledge domain of VMM (Dom0). These IDses can be controlled, configured, and monitored by cloud admin. These exists a semantic gap between VMM and VM because of having different kernel versions. Hence, some VM introspection libraries have been developed by researchers such as LibVMI [164], XenAccess[165], and DRAKVUF[166], etc. to extract the VM-related logs from VMM. These libraries are hypervisor dependent. The collected logs are passed to security tool, deployed at VMM for further analysis. In case of Xen hypervisor, the high-level information of VM programs can be extracted from privileged domain. As the security tool is working from outside, the visibility of the activities inside monitored machine is very moderate. However, the robustness and high attack-resistant of VMM-based IDSes are better than TVM-based IDSes. However, as there is a significant overhead involved during VM log collection from outside, the performance of VMM-based IDSes is also average than other types of IDSes. The VMwatcher

DOI: 10.1201/9781003004486-5

[167], VMGuard[168], and VAED[169], etc. are some examples of hypervisor-based IDS tools.

Network-based IDS performs the network traffic monitoring and are independent of underlying operating system. This is flexible to be deployed at any layer (TVM/VMM/Network). Network IDes, deployed at gateway points of cloud physical network provides primary defense from network attacks. However, the visibility of monitored VMs is not good in such cases. The performance of such IDSes in detecting VM-specific network attacks is not as good as other types of IDSes. Infact, the VM-specific attacks on OS and application programs are difficult to be detected by such IDSes. The VM program-specific information cannot be extracted from network points. Network attacks, such as Denial of Service (DoS), scanning and remote user attacks, etc., can be effectively detected by such IDSes. These tools are robust and have high attack resistance since they can't be accessed by tenants or outsides. These are hypervisor independent and are suitable to de deployed in various cloud deployment scenarios. However, network-based IDS deployment at VMM/network server is well suited to IaaS cloud and is configured by cloud administrator on demand of tenant user. SNORT-based Cloud-IDS [170] and Cloud-NIDS [171] are some examples of network-IDS.

Distributed IDS consists of multiple IDS instances of various types (as discussed above), which are distributed over the large network of cloud. These instances either communicate with each other or are centrally controlled by cloud administrator. Distributed IDS inherits the qualities of the IDS instances (TVM-based/Network-based/Hypervisor-based) deployed at the different regions. The visibility and attack resistance depends entirely on the type of IDS instances deployed, as discussed above and shown in Table 5.1. However, the performance of distributed IDS is better than other type of IDS as discussed above. Its dependency on hypervisor depends on deployment position of the security tools. Gupta et al. [172] proposed distributed architecture where Guest-IDS are deployed at each TVM (DTVM-IDS) and centrally controlled by cloud administrator. It is hypervisor independent. The Collabra [173] is a distributed architecture (DVMM-IDS) in which hypervisor-based IDS instances are deployed at each VMM and communicate with each other to update about attacks. This makes it VMM dependent. In the case of distributed IDS in which Network-based IDS instances are deployed at network points and communicate through network channels, making them hypervisor independent. For example, A distributed security framework is proposed by Lo et al. [174] (DNetwork-IDS) which is called cooperative intrusion detection framework. Multiple Network-based IDS instances are deployed in each cloud server which cooperate with rest of the instances for detecting attacks.

Each of the key IDS proposals under each of the category with the details of their detection mechanism have been discussed in subsequent sections.

5.2 TVM-based Intrusion Detection System

TVM-based IDS analyzes the specific actions of the guest by monitoring the interaction between user/system applications and guest operating system. The existing

TABLE 5.1: Types of IDSes in cloud

Parameter	TVM-based IDS (A)	Hypervisor-based IDS (B)	Network-based IDS(C)	Distributed-IDS (D)
Placement of IDS	TVM	VMM	virtual/physical network points (TVM/VMM/Network)	TVM, VMM or network points, or CCS
Visibility	High	Moderate	Low	Depends on placement
Throughput	High	Moderate	Low	Depends on placement
Resistance	Low	High	High	Depends on placement
VMM dependant	No	Yes	No	Depends on placement
Managed by	Customers	Cloud Admin	Customers/Cloud Admin	Cloud Admin
Introspection	NA	Applicable	NA	Applicable for DVMM
Tools used	BOS [162], SIM [163]	XenIDS [175] VMwatcher [167]	SNORT-IDS [170]	Collabra [173], ISCS [172] Cooperative-agent [174]

traditional security solutions for host monitoring are applicable to traditional physical server and guest machines (TVMs) at SaaS/PaaS/IaaS cloud environment. Some of them have been applied by researchers for cloud environment at TVM-layer. They cannot be directly adopted at VMM-layer because of some technical issues associated with different layers, discussed in Section 2.4. The details of some of these security proposals for TVM monitoring are discussed below.

Initially, a specification-based defensive mechanism is proposed by researhers [176]. The specification of program, pre-defines the intended behavior of the monitored programs in form of some specification policies. For example, the `rdist` is used to keep the consistency of the files which are maintained at distributed servers . This program can be evaded by following ways: The `rdist` program can be invoked by user when he/she wants to update one of his/her files running on some local host. Immediately afterward, a temporary file is created by the program before copying. The temporary file can be renamed by attacker. A symbolic link is created which is having the same name of temporary file. On completion of copying process, when the permission of the temporary file is changed by `rdist` using chmod command; it will also happen that the ownership of symbolic link is changed. It allows an attacker to take access of the temporary file. The behavior of each monitored program is identified by the specification policies. The program policies are precise. However, they require strong technical expertise.

The above mentioned approach may not be suitable for cloud. The overhead of creating specific policies for all administrative program/privilege programs running on different servers, will be quite high. In fact, it is very tedious and time-consuming task to write such policies.

Enumeration-based security approaches have also been proposed by researchers for doing attack detection by carrying out program behavior analysis. Initially, Forrest et al. [177] proposed look-ahead pair method. In their approach, a baseline db is maintained which contains two values in a pair: a system call and immediate system call sequence followed by it. The unique entries are recorded using some window size n. Each time window is shifted by one position and next look-ahead pair entry is recorded. At the time of testing, if a sub-sequence of a trace does not matches with any of the baseline db entries, it is considered to be malicious.

In the further extended work [178], STIDE approach is proposed. In STIDE, they found that look-ahead pair approach has lesser discriminating power than short patterns. A fixed size sliding window is used to generate short sequences called n-grams. A huge database is created having many short sequences of system calls as baseline profile. Any mismatch with the baseline profile signals the suspicious behavior of the programs. This approach is very older and requires huge storage to maintain the normal sequences of n-grams. It generates the false alarms for new evolving behavior. Warrender et al. [179] extended the STIDE mechanism and added a frequency threshold to it, called at T-STIDE . It not only checks the matches of incoming sequences with the baseline profile but also matches the frequency of occurrences with some threshold value.

Motivated from the frequency-based approaches, some researchers presented the use of traditional 'Bag of system calls (BoSc)' [180] in a cloud environment. For example, Alarifi and Wolthusen [162] used BoSc to detect the malicious activities in VMs. Each VM runs an IDS instance and captures the system call logs and performs analysis to identify the suspicious activities inside VM. The IOCTL system calls are collected over a period of time in KVM server using the *strace* utility of Linux. It achieves a 100% detection rate and false alarms of 11.11%. The major limitation with the frequency-based approaches is that they lack in maintaining the order of patterns.

Yin et al. [181] employed Hidden Markov Model (HMM) for detecting the anomalies. It was tedious and time-consuming to build the specification profile of each and every monitored program. There is a stability in using temporal signatures and hence building profile using them would enhance the efficiency of IDS. The authors have used HMM for building profile of such behavior of all the normal sequences. However, the training time of the model is quite high. The evolving new behavior of the traces can also be classified as suspicious.

The state-based approaches are employed by Alarifi and Wolthusen [182]. The normal behavior of the programs is modeled by using Hidden Markov Model (HMM). The normal behavior of the programs is captured and is used to build normal profile. The approach is deployed in the KVM-based hypervisor. HMM generates the state diagram by modeling the system call sequences of normal programs. Each transition of state diagram is denoted by a probability value of going from one system call to another system call. The model has been validated over some self-generated DoS attack dataset. In the first scenario (all types of system calls are considered), it provides 100% detection rate with a false alarm of 5.6%. In the second scenario (IOCTL system calls are considered), it provides 83% detection rate. In cloud environment, the normal behavior keeps on evolving. Hence, retraining a model is very costly and time consuming. Secondly, the knowledge of having the prior information about states is not feasible when it comes to modeling the dynamic analysis-based approaches.

VMAnalyzer [183] is one the efficient TVM-based malware detection framework cloud which makes use of deep learning approach as shown in Figure 5.1. The security tool runs inside the monitored VM where the user programs are running. It uses the guest OS utilities such as *strace* to extract the long sequence of system calls. It makes use of convolutional neural network (CNN) for learning the long ordered sequences. In layer 2, it employs Long Short-Term Memory (LSTM), to learn and detect the behavior of malicious system call sequences. The approach is validated using a standard dataset and results seem to be promising.

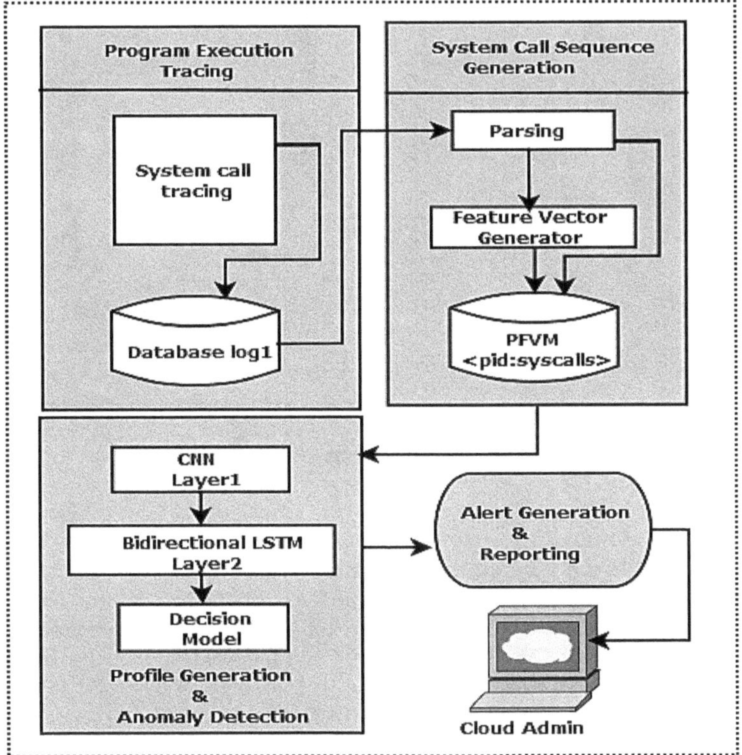

FIGURE 5.1: VMAnalyzer: basic cloud security architecture.

5.3 Hypervisor-based Intrusion Detection System

Hypervisor-based IDS provides a strong security by analyzing the VM logs from outside. It makes use of introspection libraries to extract the VM state information from outside. Since, the IDSes are placed in a trusted domain, it becomes difficult to breach such security tools. Each of the IDS instance runs individually on hypervisor/VMM in the VM-hosted cloud servers. They are controlled by the cloud administrator. It is easier for VMM-based approaches to detect the stealthy attacks that try to hide their behavior from the VM security tools. These solutions are more applicable to IaaS cloud infrastructure. Let us now discuss some of such approaches:

Livewire [184] is a VMI-IDS architecture, proposed by Garfinkel as shown in Figure 5.2. It is deployed in the VMM-layer of VMware workstation which makes it more robust to attacks. The communication between VMM and VMI-IDS happens through a command line interface (CLI). The CLI is used to configure, introspect or monitor the security policies. The two key components are: OS interface library and policy engine. A detailed information of all the VM states is provided by OS interface library. The events generated from these two components are notified to the policy engine. The policy engine makes use of signature-matching technique to detect the anomalous activities in VM. It fails to detect the unseen malware activities.

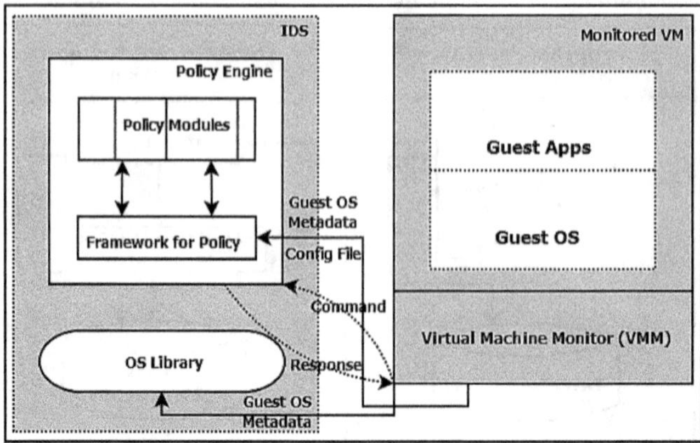

FIGURE 5.2: Virtual machine introspection-based IDS architecture.

The guest view casting technique is implemented as a prototype by VMWatcher [167]. The VM memory information is extracted from VMM. The mechanism is OS depended and requires the knowledge of the guest OS architecture to be monitored. For example, for Linux, the guest OS symbol details can be provided `System.map` file, which is provided by OS distributor. For windows, the memory is completely scanned using specific signatures (such as 0x03001b0000000000). It helps in identifying the potential attacking processes in the memory. The approach is helpful in detecting hidden processes. However, it does not perform a detailed behavior analysis.

A malware severity analysis mechanism [185] also proposed by some researchers for cloud. The security monitor is deployed at VMM. Various security requirements created for the model are as follow: integrity of guest OS, integrity of work state, zombie protection, DoS, resource exhaustion by malicious software, backdoor security, platform attacks. REMUS ([186]) classification is used to map the security requirements and system calls. Once a system call is found suspicious, it is then analyzed by the severity analyzer which employs a machine learning classifier (DT C4.5) in order to classify system calls. A detection rate of 90.7954% is achieved for a self-generated dataset. Although the security mechanism does not give any clue about how system calls are extracted.

5.4 Network-based Intrusion Detection System

The network-based IDSes captures the traffic from physical/virtual network interfaces. They detect the network attacks such as scanning, spoofing, DoS attacks, etc. These approaches do not require any change in the hypervisor which makes them portable. These IDSes can be deployed at any layer such as TVM, VMM or network. However, ideally, the network points such as gateway, network servers, etc.

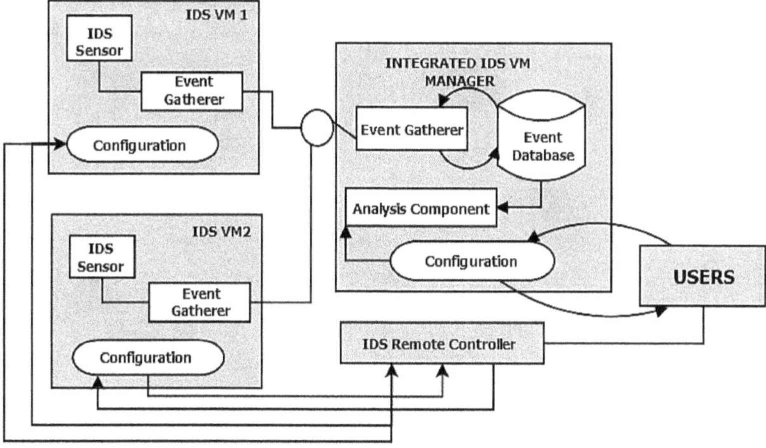

FIGURE 5.3: VM-integrated IDS.

are prime locations where they are usually deployed. Let us now discuss some of the NIDSes mechanism.

Roschke et al. [187] proposed a VM-integrated intrusion detection architecture that employs signature matching technique as shown in Figure 5.3. There are various components of their architecture: event database, event gatherer unit, analysis component and intrusion remote controller. There are deployed in a central server which collects the events from individually VM. Each of the monitored VM runs an IDS sensor. The sensor generates the log messages in certain format which are exchanged with analysis server. These messages are correlated to identify the distributed attacks . A remote controller is provided to configure the sensors as part of central unit. The techniques fail to detect completely unseen attacks. A centralized NIDS architecture [188] is proposed in past research for detecting cloud-based network attacks. A NIDS tool (Snort) is installed near to cluster control server to identify the flooding attacks. The central server does all analysis on the network traffic. The experimentation is performed in the Eucalyptus test bed set up. The architecture is prone to single point of failure.

Another past research explains a network attack detection mechanism [189]. In this mechanism, a anomaly score is created for various security groups. An according to the score (severity), the security policies are designed. Whenever a request is received from the user, an Accounting, Authorization and Accounting (AAA) module is invoked. The AAA module generates a anomaly score of user and accordingly appropriate IDS is selected for the specific user VM instance. The user information such as transaction logs are kept in certain databases which are communicated to AAA module periodically. There are three types of security levels defined: high, medium and low. The high level defines the strong security polices with IDSes. Medium level defines moderate and low level defines baseline security policies with IDS. All the security rules follow signature matching mechanism and hence require regular updation of the signature database.

The eCloudIDS [190] is another security framework, designed to detect malware in the VMs running in cloud. The following key design components of the system

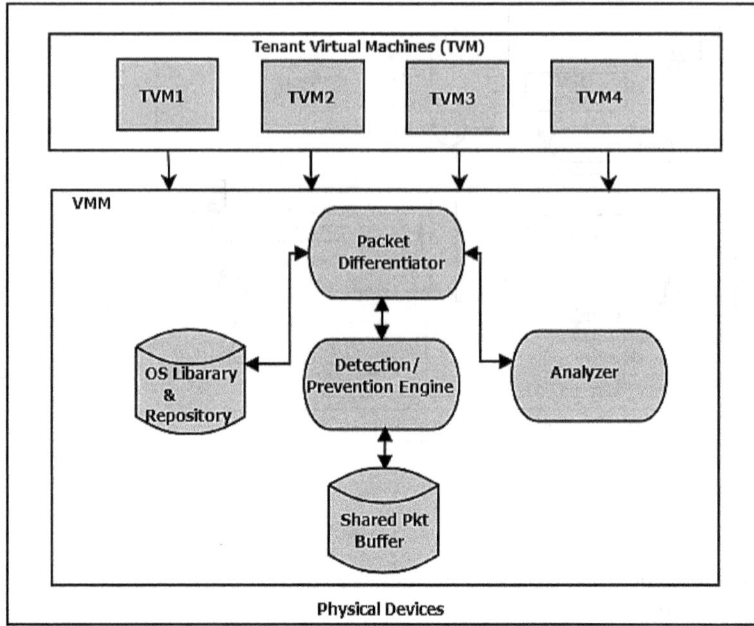

FIGURE 5.4: VICTOR intrusion detection system architecture.

are: uXEngine and sXEngine. The former uses unsupervised machine learning (self organizing map) whereas later component is based on supervised machine learning technique. The actions performed by the user and their application logs are logged in H-log-H component. The log and pre-processor extracts the user processes and the audit logs, taken from H-log-H. It makes it compatible for the uXEngine. Then later the logs are processed by pre-trained sXEngine model and classified in a specific class. It achieves 89% detection rate and 2% false alarms. Modi et al. [191] have done significant amount of research work in cloud-based NIDS. In one of their proposed NIDS frameworks, the primary level of detection is done by a signature matching module. If the VM traffic pattern matches with any of the defined signature, alert signals are generated. If the traffic is found to be normal then again it is passed to a second level of check in which a pre-trained anomaly detection module is used to decide whether it contains an abnormal traffic flow. Again if traffic is characterized as suspicious, an alert is generated to cloud admin otherwise no action is taken. The Snort is used as a primary detection module whereas decision tree is used as secondary decision module. They have performed validation using KDD'99 and NSL-KDD'99 dataset. It achieves the 96.71% detection rate with 4.81% false alarms using later dataset.

VICTOR [192] is an IDS architecture for IaaS-Cloud, proposed by Tupakula et al. for detecting attacks as shown in Figure 5.4. It consists of various security design modules such as OS library and repository, packet differentiator (OSR), analyzer and packer buffer and intrusion detection engine (IDE) . The model is deployed on VMM or host OS. VM packets are received by packet differentiator. The details of VMs, processes, OSes and applications are updated in the OSR. It verifies and checks for the verification of the information that is reported by packer differentiator. If any of the hidden process is detected, an alert is raised to IDE. The suspicious packets are matched against known attack signature database. Afterward, anomaly module which is based on machine learning updates the OSR for the new behavior of the VM. The legitimate packets are passed to the destination machine. The framework is useful to detect attacks from hypervisor and suitable in various cloud IaaS scenarios.

A VMM-based network security framework [193] is provided in past to detect attacks in cloud environment. The network traffic traces are collected at the VMM in various benign and intrusive scenarios. The proposed approach first of all, creates the clusters of data by making use of fuzzy c-means clustering technique. The clustering helps to group the similar datapoint together. It also helps to create the sub-set of dataset which are further passed to ANN classifiers in a parallel fashion. The sub-set of dataset is given as an input to each of the ANN algo. The results is further combined with the fuzzy-aggregation module. The technique has been validated using KDD'99 dataset. It provides a detection rate of 97.55% and 3.77% false alarms. Although the training time is comparatively higher than other mechanisms.

Another novel security framework [194] integrates the firewall with intrusion detection system. Initially, firewall filters the traffic as per the rules defined prior. The blocking rules are also defined as this level. Once the first filtering is passed, the traffic is further passes through the second component which is intrusion detection system. There are various attacking rules defined in the second component as well which does the deep signature scanning and tries to find out if the traffic contains any suspicious activity. If the traffic is found as suspicious, then the firewall security policies are updated accordingly. The system performs efficiently to identify some self-generated attacking traces of DDoS attack.

A novel network attack detection framework [195] is proposed based on Openstack cloud. The NIDS is proposed as a service to provision on demand tenant traffic monitoring and analysis. Tenant has flexibility to delete or create the NIDS service instance as per the need of applications. The attack rule database can be updated on request by making use of this service. Instead of running the NIDS instances on individually host, the service runs on a dedicated host. It reduces the virtualization overhead and supports light weighted and quick deployment of NIDS service. The prototype has been implemented using multimode openstack set up.

5.5 Distributed Intrusion Detection System

In case of distributed IDS systems, the IDS instances are distributed at various locations (VM, or VMM or Network) in cloud environment. These IDSes communicate to each other or some central monitoring server with the alerting logs and reports. This kind of architecture is very helpful in collectively analyzing the attack logs to identify distributed attacks.

Che et al. [196] proposed distributed architecture which is based on state transition analysis approach for detecting intrusions in cloud. The analysis component executes in each of the monitored machine and analyzes the logs of user activities. The user activity logs are important data files which hold the traces of all types of user behavior. The attack patterns may also very from attack to attack. The attack correlation module is then used to correlate various user activities to identify the anomalous activity present in logs. This analysis is helpful in detecting specific attack patterns. The approach uses the signature matching technique and is helpful in identifying the known attack behaviors. The attack traces which do not match to any signature pattern can also be classified as normal.

Gupta and Kumar [197] proposed a distributed dynamic analysis-based intrusion detection framework to detect attacks in cloud. In their approach, IDS instances are distributed across various VMs in the cloud network for doing the VM program analysis on monitored VMs. These instances reports to the centralized server which is configured to take appropriate actions. The approach uses a pattern matching approach to detect the suspicious system call pattern in the monitored programs. If a pattern does not to any baseline signature, an alert is raised and reported to central administrator. The program-wide detection is helpful in detecting the program abnormalities induced by attackers. The approach achieves a detection rate of 98% in detecting intrusions. It lacks in employing any kind of machine learning mechanism which limits is capability in detecting new evolving program behavior.

Varadharajan et al. [198] designed an integrated IDS framework for cloud environment. The framework employs the attestation techniques and access control policies as primary security measures. The analyzer components are deployed at the privileged domain of hypervisor of each of the VM-hosted server. A trusted platform module is deployed at the physical server is hosting the user VMs. The hypervisor layer is integrated with the intrusion detection engine, decision making engine and access control component. The intrusion detection module helps in identifying the anomalies. The access control defines the access control policies for the monitored VMs and decision module takes the security decisions what to do in response to the on detection of some suspicious behavior of the system.

In addition to the above mentioned distributed frameworks, some of the distributed frameworks are based on the collaborative mechanism in which the security instances are distributed at various locations in cloud. They share the security related event information with each other to identify the distributed attacks.

Lo et al. [174] proposed a cooperating agent-based distributed intrusion detection framework to detect the attacks in cloud. The Snort IDS is deployed at various cloud servers which is integrated with various other security modules such as threshold computation, alert clustering, intrusion response and blocking operations. Whenever, Snort detects any suspicious activity in the system, the next module alert clustering module is invoked. The severity of the flow is identified by this module based on certain threshold value. The alert response invokes the blocking module once the threshold crosses certain value. The blocking module at each monitored server helps in adding the blocking rule as response to collective decision taken by the servers on detection of some suspicious flow. As the approach uses the signature matching mechanism; it cannot detect completely unseen attacks.

There exists a distributed VMM-based architecture [173], used to detect the distribute attacks in cloud. It is a VMM-based framework called as collabra, designed to detect attacks at the most trusted region of cloud. The two main components of collabra are: hypercall integrity checker and hypercall origin checker. The integrity checker checks if the hypercall is maliciously modified or not. It verifies the message authentication code (MAC) of the hypercall from repository along with the specified policy. For unknown calls, the origin is again checked and applications and

VMs with their privileges are identified. On detection of suspicious hypercall, an alert information with details of calls are communicated with the other collabra instances running in other VMMs through the logical channels. The prototype is not implemented by authors.

A autonomous hierarchial framework is proposed by Kholidy et al. [199] to detect the attacks back end cloud servers. The security instance is deployed at the VM management server. The events sent by sensors are collected by the collector module. The event analyzer (both NIDS and HIDS) module analyzes the events, correlates the events and detects the suspicious user activities across several VMs. The suspicious events are communicated to the controller module to take timely actions. An alarm is send with the risk factors to all other controllers deployed in cloud environment. The approach is suitable to detect various basic attacks. However, stealthy attacks which target hypervisor or hardware layer cannot be detected by such traditional security mechanisms.

A collaborative intrusion detection framework collaborative intrusion detection framework [200] is proposed by Haddad et al. to detect cloud attacks. A security tool (Snort IDS) is deployed in each of the VM hosted server to detect the network-level attacks. The IDS instances matche the abnormal flow in the system by matching the traffic patterns against a pre-defined attack signatures. On detection of suspicious alarm, the alert signal is communicated to the cloud administrator. If the IDS instance passes some traffic and marks it normal, a second level of filtering is applied to the traffic. In second stage of filtering, SVM classifier is used to find the abnormality in the traffic flow. It classifies the traffic based on the learned attack patterns. In the second stage, the attack variants which are by-passed in first stage are detected successfully. The information about the attacks and signatures is communicated to the other IDS sensors and signature database is updated.

5.6 Research Challenges

There are various security concerns in cloud as discussed in the previous sections. On the basis of critical literature review on intrusion detection in cloud, some research challenges have been identified. Let us discuss each of them in detail:

Detection of Malware Attacks

Signature-matching and static analysis techniques can be subjected to code obfuscation attacks. Dynamic analysis overcomes this limitation. The existing system call analysis approaches for cloud such as Bag of System calls (BoS) [162, 201] and Immediate System Call Sequence Detection (ISCS) [172, 197] have some limitations associated with them. In BoS, system call ordering is lost which may be a weakness of such techniques. Though, ISCS retains the ordering, it fails to capture the behavior of very longer traces. Furthermore, these approaches assume to have a full knowledge of the normal execution behavior of programs. In addition, they do not apply the statistical-learning-based techniques IDS approaches [197], which are very important for generalizing the program behavior.

Subversion of Security Monitor

An attacker can check the presence of security processes running at TVM and can try to disable it. For example, malware such as Torpig and Conficker disable the security tools and other security-critical services such as auto-update, error reporting and Windows Defender services [156]. Some of the malware hide their presence from the security monitor running at TVM. In addition, advanced malware can refrain from executing if they detect the presence of security components at the TVM. The existing IDS approaches [182, 197, 201, 202] do not consider such evasive behavior. Hence, a CSP must provide a more robust defensive solution to detect stealth malware at VMM-layer.

Detection of Network Attacks within Virtualization-Layer

Hypervisor supports the virtual networking to allow the communication between the co-located VMs. The traditional NIDS and existing frameworks, that deploy IDS at central controller/cluster node/cloud physical server [170, 200, 203, 204] do not perform the fine granular detection at virtualization layer. It is difficult to track or identify the malicious VM from outside. The situation becomes more complex if the attack packets are forged by virtual IP/MAC address. Most of the cloud security frameworks apply traditional signature-based NIDS at TVM-layer to detect network intrusions [171, 187, 189, 190]. Regular updates are required in case of pattern matching approaches. In addition, applying the signature matching at VMM-layer [205] [206] may impose a significant overhead to the system. It is because of the time complexity that is associated with the rule-matching mechanism that will lead to high computational cost at the time of detection. Some recent works [193] apply machine learning for detecting network intrusion at VMM-layer. However, they are not integrated with the network introspection functions.

Virtual Machine Introspection (VMI) Enabled Security

We identified that not much work has been done in the direction to provide VMI-based security solutions for cloud. The existing security solutions have some limitations associated with them which make them less suitable in a multi-tenant cloud environment. For example, some approaches [207] impose restrictions such as the monitoring VM and monitored VM should have same OS installed. ShadowContext [208] is an attack prevention strategy based on system call redirection approach. An improperly configured system call redirection module can crash the operating system kernel. Some of the introspection approaches [167] provide the limited introspection functions and are not sufficient for intrusion detection applications. Xenini-IDS [175] is proposed for virtualization environment which does the behavior analysis using very older STIDE approach [178]. STIDE makes use of string features and is prone to string manipulation attacks. Moreover, its detection mechanism only rely on all normal sequences which will produce more false alarms in dynamic cloud environment. Therefore, there is a need to provide a more efficient, VMI enabled security framework which is compatible to work for cloud deployment and is integrated with efficient detection mechanism.

Feasibility of Same Security Solution at All Layers

Distributed frameworks overcome some drawbacks of existing IDS frameworks [180, 182, 196, 202, 208, 209] that were not centrally controlled and hence prone to attacks from nefarious users. However, it is being identified that distributed TVM-layer approaches cannot be applied directly at VMM-layer. This is because Hypervisor/VMM can view the guest operating system's information as raw bits and bytes. At the VMM-layer, high-level semantics of TVM is not known. For example, the information about a TVM such as processes, data structures, files and OS abstraction, etc. are not recognized from Hypervisor. In the same way, distributed network-layer approaches cannot view the guest-specific information from outside. Hence, the practicability of the security frameworks [174, 191, 197, 200, 203, 210], which claim to deploy the same security solution at all regions in cloud sounds less feasible and inefficient.

Lack of Efficient Security Architecture

Majority of IDS proposals for cloud [171, 193, 205, 209, 199] analyze the traffic patterns to classify the network flows. It can be helpful to detect network attacks. However, low-frequency attacks such as virus, worm may not be detected efficiently by merely having such analysis. Some of the researchers [197, 201] worked in the direction of process analysis at the TVM-layer of cloud. However, again only process analysis is not sufficient for detecting network attacks. Hence, a security architecture which performs both traffic analysis and system analysis is required.

Robust Layered-Security Architecture

Distributed IDS which are centrally controlled by cloud administrator or which cooperate with each other for attack detection are robust than other types of IDS. However, the existing distributed security architectures support distributed guest-based IDS [197] or distributed VMM-based IDS [173] or distributed network-based IDS [174, 203]. These IDSes detect the attacks at specific-layer. There is a need to provide a more robust security architecture which covers all three-layers of cloud, i.e., TVM-layer, VMM-layer and network-layer, addressing the design limitations at individual layer to facilitate the detection of both malware and network detection in cloud.

Inefficient Conventional IDS Solutions

There exist some cloud security frameworks that employ the conventional IDS frameworks [191, 171], which make use of traditional pattern matching-based tools such as Snort and integrate the output with some traditional machine learning-based IDS in serial order. The use of signature matching and anomaly module may improve the detection rate for known attacks and attack variants. However, zero-day attacks are still difficult to be detected. Moreover, the solution is not virtualization-specific and can also be applied in traditional security environment.

Less Solutions for Improving Hypervisor Security

The virtualization specific attacks can target the hypervisor and manipulate the hypervisor-level kernel structure. Infact, some attacks like VM Escape tries to take the control of hypervisor root terminal from guest domain by running some malicious scripts. Hence, merely assuming that hypervisor is always safe may not work always. A strong hypervisor security mechanims should also be in place when we talk about cloud security.

We can conclude that there is need to provide a more robust and efficient security architecture to deal with major security challenges identified.

5.7 Conclusion

The categorization of key IDS security proposals have been provided with discussion of their pros and cons. The detailed analysis of techniques provides readers a strong and coherent view on security solutions proposed so far. We can conclude that there are various types of IDSes in which researchers are working on. Traditional tools can be used as TVM-based IDSes. VMM-based IDSes are more specific to Cloud Virtualization layer and are designed based on the hypervisor design. VMM-IDSes are more attack resistant and robust. They are more successful in revealing the stealth behavior of malware than traditional IDSes. Network-based IDSes are more specialized to detect network attacks in cloud and are more appropriate to be deployed at Cloud network gateway points or network servers. The distributed IDSes are more advanced IDSes, which perform the attacks correlation and identification to detect the distributed attacks.

5.8 Questions

Fill in the blanks

1. Mark the correct statement. The throughput in TVM-based IDS is

   i High
   ii Moderate
   iii Low
   iv None of these

2. The tools/architecture for hypervisor-based IDS

   i Snort
   ii Cooperative-agent
   iii VMWatcher
   iv VMGuard

3. Network-based IDS is managed by

   i Cloud administrator
   ii Tenant administrator
   iii Cloud customer
   iv both i & ii

4. Mark the approaches to detect malware attack

   i Bag of System calls (BoS)
   ii Immediate system call sequence detection
   iii both i & ii
   iv None of the above

5. Mark the correct statement

   i Network-based IDS deployment at VMM/network server is suited to IaaS
   ii The visibility and attack resistance in distributed IDS does not depend on the type of IDS instances
   iii The hypervisor-based IDS is not VMM dependent.
   iv The visibility in Network-based IDS is high.

Short-Answer Questions

1. Describe the TVM-based IDSes with characteristics and working mechanisms.
2. Explain the difference between TVM-based IDSes and Network-based IDSes.
3. What are Network IDSes? Exlain their working with limitations

Long-Answer Questions

1. Explain the deployment scenarios of various types of IDSes in a comparative way.
2. Explain ShadowContext mechanism in detail with suitable diagram.

Chapter 6

Intrusion Detection Techniques in Cloud

6.1 Introduction

Intrusion detection systems (IDSes) are the most powerful defensive solutions, used to deal with attack threats in the emerging era of computing. Cloud-IDSes are the defensive tools which are used to capture the cloud network traffic or/and host/VM system logs and to analyze the system or network for the presence of any suspicious activities on detection of malicious flow or pattern, an alert is sent to the admin and/or tenants.

The various intrusion detection techniques have been categorized into following types: (i) Misuse-based detection, (ii) Anomaly-based detection, (iii) Introspection, (iv) Hybrid technique, as shown in Figure 6.1 [144]. Misuse-based detection techniques are capable to detect the known attacks and they maintain their signatures in the attack database. The signature-rules for known suspicious behavior are derived by using knowledge-based systems or machine learning algorithms. Anomaly-based detection techniques try to learn the existing ordinary behavior of the system over some time period. If they found any mismatch from the learned behavior, a notification for suspicious activities is generated and sent to admin [211].

The above detection approaches cannot be applied directly at the hypervisor because of difficulty in interpreting VM semantics at lower layer. This semantic gap refers to interpreting the low-level information of guest OS into a high level semantics at the hypervisor. Introspection techniques are advanced virtualziation-specific techniques which help in resolving the gap and providing the information about the VM memory from VMM or other monitoring VM. It analyses the VM code running inside VM and extracts the behavioral semantics. The program activities are analyzed for the presence of any malicious code by incorporating security analyzer with the introspection libraries [212]. Hybrid approaches combine the above explained techniques for producing better results.

In the beginning, the traditional IDSes have been used in cloud environment. For example, Snort-based security framework is proposed by Roschke et al. [187], called as VM-integrated IDS to detect the network anomalies in cloud. Some authors [191] proposed a security framework in which machine learning classifiers are integrated with Snort for detecting network attacks in cloud. Some authors [162] used 'Bag of System Calls', a traditional approach, to detect the suspicious system call patterns in the execution traces of monitored programs. Furthermore, a dynamic analysis-based security approach, called immediate sequence of system call approach is introduced

DOI: 10.1201/9781003004486-6

[197]. In that approach, a feature vector in form of <key, value> pair is generated for each monitored program.

Li et al. [209] employed ANN to learn the network traffic behavior of normal and anomalous connections. Pandeeswari and Kumar [193] proposed an security framework which combines fuzzy logic with ANN to detect attacks at hypervisor layer. In all the above mentioned IDS frameworks, deployment of security tool is done at the cloud servers or VMs. However, while adopting the traditional IDSes, the access privileges, security roles and deployment locations need to be carefully designed.

Moreover, the traditional signature matching or static analysis-based IDSes are likely to be subverted by attackers by incorporating anti-malware detection techniques such as obfuscation and encryption [213]. A malware program can thwart an IDS running in the monitored VM by various ways. Some of the malware terminate the security-related processes such which make the system unsafe from recent malware attacks. Some of the advanced malware tries to sense the virtual environment by scanning through system registry and looking for the presence of specific drivers to virtual environment. On sensing the virtual environment, a malware may behave differently and tries to bypass the security analysis tool running inside the virtual environment [169].

These advanced malware are called as evasive malware. There are some rootkit malware which tries to modify the guest kernel of the VM. These malware hide themselves from the security tool, running inside the tenant VM. It is crucial to detect such malware in an early stage so that the advanced attacks such as side channel attacks, VM escape attack, etc. can be minimized to a greater extent. Traditional HIDS are not capable to detect such advanced attacks. Advanced cloud IDS system based on memory introspection such as vProval [214], Maitland [202] and VMGuard [168], etc. detect such malware by taking the VM memory view from outside.

Some of the researchers have proposed traditional network IDS [187] [170] which makes use of rule-based signature database with intrusion detection mechanism. The major limitation with rule or signature-based approaches is that they require regular maintenance of the attack database and fail to detect unknown attacks. The extended work of Modi et al. [171] integrates the signature matching with machine learning approaches. The anomaly module uses machine learning algorithm and is executed after signature matching module. Most of the IDSes have been validated with KDD99 dataset [157] which is very older dataset and does not have the cloud network traffic statistics. Hence, it is challenging to assess the overall efficiency of the system in real environment. In fact, traditional NIDS, deployed in cloud server or network points, fails to identify the attacks targeted from one VM to co-located VMs. The traditional tools work efficiently in physical environment and less suitable in handling virtualization-specific threats.

There are some advanced introspection techniques such as Virtual Machine Introspection (VMI) and Hypervisor Introspection (HI). VMI approaches are used to extract the VM information from hypervisor. HI techniques are used to introspect the hypervisor, extract and analyze the details of hypervisor. These techniques are helpful to detect virtualization-specific attacks such as cross-VMside channel attacks and VM Escape attacks, etc. HI techniques are helpful to detect hyperjacking attacks and VMM backdoor attacks, etc which are undetected by traditional IDSes. There is a huge gap between introspection-based security tool and attacker.

A high confidence barrier is provided by the Introspection-based approaches between attacker and security tool when compared to traditional IDS. CloudVisor [215] is one of the HI-based security analyzers which is based on nested virtualization and is used to detect hypervisor-level attacks. Hypercoffer [216] is also an introspection-based IDS which maintains the integrity of the tenant VMs. A hybrid architecture is proposed by Varadharajan et al. [217] which combines the introspection with anomaly and misuse detection approaches and provides improved detection accuracy. VMI based approaches assume that hypervisor or VMM is a trusted domain. However, HI based approaches do not assume such facts and hence more suitable for hypervisor security.

6.2 Taxonomy of IDS Techniques

IDS techniques can be classified into various subcategories: (i) Misuse detection, (ii) Anomaly detection, (iii) VMI-based techniques, (iv) HI-based techniques, and (v) Hybrid techniques, as shown in Figure 6.1 [144]. These techniques are further subcategorized based on detection approach. Each of the techniques is described in detail in below sections.

6.2.1 Misuse detection techniques

In case of misuse detection techniques, an attack database of prior known attack is maintained. The current behavior of the system is matched against the known pre-defined attack patterns. A conceptual diagram for knowledge based approaches is given in Figure 6.2 [144]. In cloud, these approaches are deployed at the tenant VMs. They can detect the VM-level attacks. Cloud-based IDS techniques under this category are mainly of two types: Knowledge-based and machine learning-based, as discussed below:

A. Knowledge-based

In this category of approaches, VM program logs/traffic logs are compared against pre-defined attack patterns. These approaches are classified into following sub categories: (i) Rule-based expert systems, (ii) State transition analysis, (iii) Signature analysis [218]. The expert system (rule-based) maintains a database of rules for attacks in various malicious scenarios. Signature-based approaches maintains the database of attack patterns or signatures. The incoming traffic packets are matched against these attack patterns. A transition model of states is maintained in case of transition analysis approaches. The various paths of the transition model may lead to a compromised or non-comprised system state.

Some researchers have proposed signature-based techniques for detecting attacks in cloud. A VM-integrated IDS is proposed by Roschke et al. [187], where VM-IDS instances are deployed inside VMs. A management node manages all these IDS instances. The management node performs the event gathering and maintains a event database. It performs the analysis on the collected log and controls the IDSes

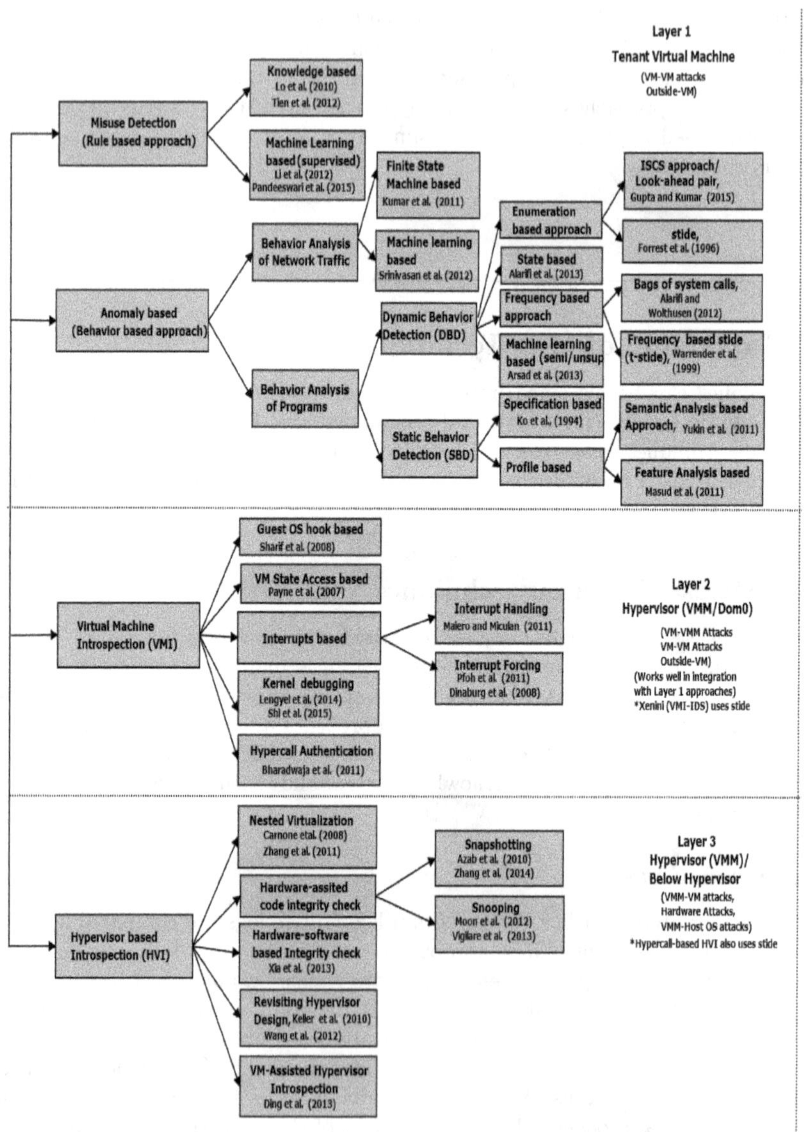

FIGURE 6.1: IDS techniques: a taxonomy in cloud.

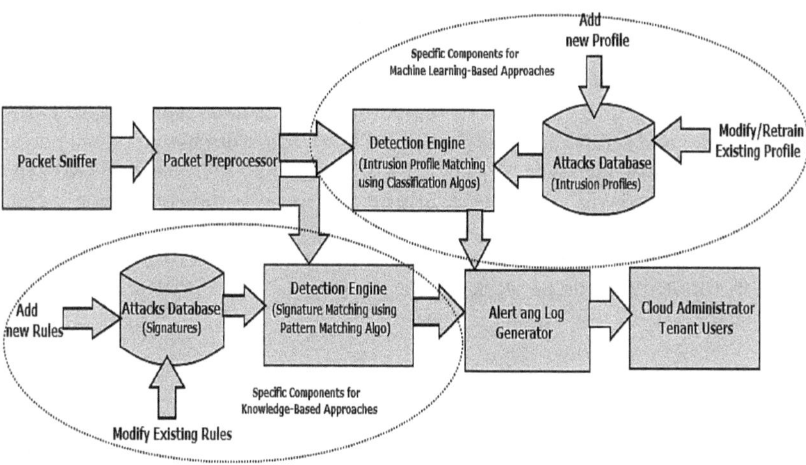

FIGURE 6.2: Conceptual working of misuse detection approaches.

running in VMs remotely. IDS instances send alerts to the management node which then performs the format conversion and generates Intrusion Detection Message Exchange Format which are further stored in the database of events. The actions on the detected anomalous traffic can only be taken by the management node.

Another signature-based attack detection framework is proposed by Lin et al. [205] for detecting network anomalies, deployed at the secure VM or Dom0 VM. The signature-rules are created and maintained for each VM based on the applications running inside the VM. Once the VM information is extracted, the signatures are created dynamically. For example, in windows based machines, MTOM and EPROCESS data structure are referred by kernel to check the VM processes and other related information. They have used SNORT NIDS and considered Linux platform for validation.

A hierarchical cloud-based IDS is proposed by Kholidy et al. [199] which is used to secure back-end servers and VMs. There is a node maintained, called as management server which sends alerts to the controller with a risk factor value. The various components of framework are event collecting module, correlating module, event analyzing module and controlling module. The analysis module consists of NIDS, HIDS, and DDSGA analyzer. The event collecting module performs the log acquisition from host machines and network events from IDS sensors. The correlating module correlates the host and network events to check the changing behavior of users in various VMs. The statistics calculated by correlator are analyzed by the analyzing module. On detection of suspicious activity, alerts are generated and controller is informed.

Some of the researchers have applied parallel computing approaches with IDS approaches which can also be applied in cloud environment. For example, Divide Data Parallel (DVD) [219] is an IDS approach which distributes the packet payloads among processors. The algorithmic computation is distributed among processors which reduces the overall time delay. It also employs a synchronization mechanism to reduce the risk of overlapping of data. However, the approach fails to detect the attacks whose attack signature spans across the fragments.

A NIDS framework is proposed by Vokorokos et al. [220] which makes use of cluster of nodes having GPU processors. There is a master node which coordinates among these nodes and sends them instructions. The architecture has got three parts: (i) network traffic capturing layer, (ii) intrusive activity analysis layer, and (iii) synchronization managing layer. The coordinating node controls the entire monitoring environment. It maintains a database where all the alert records are acquired from various sensors. The major drawback with this approach is that it does not consider any change in topology of the network fat the time of the rule updation.

B. Machine Learning-based

There are some well-known intrusion detection techniques which are used to classify some known attacks. These algorithms generate a decision model which is a trained classifier over a well-known dataset. Decision model represents attacking profile in various attacking scenarios. These profiles are generated with the help of learning and are not predefined.

Kim el al. [221] employed SVM for learning the intrusive and normal behavior of network traffic. SVM generates the optimal hyperplane in the learning phase which is defined by support vectors and associated weight values. Many tweaks in the parameter setting has been done to find the parameters tuned with the dataset. Amiri et al. [222] provided a feature selected approach, called modified mutual information based feature selection which is integrated with SVM to detect the network anomalies. The important network features are selected and dataset is optimized. Now, SVM is trained on the optimized dataset and provide a better accuracy.

Some authors [223] presented the use of fuzzy-c means for intrusion detection. Fuzzy c-means clustering transforms the features to items. Some of the researchers have applied machine learning algorithms for securing the cloud network from network intrusions. Li et al. [209] proposed the use of BP ANN in a distributed IDS framework where IDS sensors are distributed across cloud servers. The BP-ANN algorithm is used to learn the traffic behavior of VMs both normal and intrusive scenarios and VM profiles are created as baseline database profile. On detection of malicious activity, an alarm is raised to the cloud administrator. The approach provides 99.7% accuracy with seven nodes.

Pandeeswari and Kumar [193] proposed the use of ML to detect attacks at VMM layer. The technique captures and analysis the virtual traffic collected from VMM layer in both normal and intrusive scenario. They have first used fuzzy c means clustering to create groups of input data based on the membership value. Further, they have applied ANN on each of the clusters created in earlier stage. The results produced by each of ANN instance on individual cluster are aggregated and a common decision is produced for the monitored input data stream. The approach provides 97.55% detection rate with 3.77% false alarms. It has been validated with a very older KDD'99 dataset and its real time performance, has not been validated.

The major limitations with the supervised machine learning are that they fail to detect zero day attacks (which have not been discovered). Although they are powerful in detecting known attacks and variants of attacks. Further we explain anomaly detection approaches below.

6.2.2 Anomaly detection techniques

The approaches maintain the database of expected behavior of the system over some time period and generate a model for decision making. If any deviation from the learned profile is found, the activity is signaled as suspicious. In these types of techniques, there is no knowledge about the attack signatures and no pre-defined rules exist. The basic idea of these approaches is to identify any such behavior which does not matche to normal profile. These techniques work well in detecting completely unknown attacks, called zero-day attacks (vulnerabilities which have not been identified so far). Many researchers used this type of analysis and proposed various cloud-based IDS. We have categorized the work on anomaly detection in cloud in two categories: (A) Network traffic-based analysis and (B) Program semantic-based analysis. The program semantic behavior analysis techniques have been classified into two parts: (a) Program dynamic analysis and (b) Program static analysis.

In network traffic based analysis, network sniffing tools such as tshark and tcpdump are used to capture the network traffic logs. The program dynamic analysis techniques capture the program execution traces, extracted using programs tracers such as DRAKVUF [166] and Nitro [224]. Whereas static analysis approaches captures the static features such as opcodes, API calls, DLL functions, hex codes, etc., using program disassembler such as IDA PRO [225]. The collected raw log files are first of all processed and required features are extracted using various feature extraction approaches. The pre-processed data is further passed to a behavior analysis engine. The behavior analysis engine matches the current behavior with the expected behavior and signals intrusion if mismatch is detected to the cloud administrator. These approaches can be directly deployed inside the tenant VMs of the virtualization layer or network monitor and can detect VM and network related attacks. The basic conceptual working of the anomaly detection approaches can be referred from Figure 6.3 [144]. Let us now discuss various anomaly detection methods.

A. Anomaly-based network traffic analysis

Network security-based approaches capture the ingress and outgress traffic from monitored network. The traffic characteristics such as src port no, src IP, dest. port no, dest. IP, protocol, etc. are some of the basic network flow features, which are important features used during VM profile creation. The anomaly based network intrusion detection is categorized into following types: machine learning and finite state machine (FSM)-based mechanisms [226]. The FSM-based and ML-based approaches are mainly used by researchers for cloud-based network intrusion detection systems. The description under each category is given below:

A FSM-based decision models consist of various states and transitions. The past information is stored in form of states. The change in states is represented by state transitions which may happen on occurrence of an event. Each event has certain response which is represented by actions. One of the existing state-based analysis approaches [227], which employs Hidden Markov model to generate the system behavior in form of state transitions overt some time period. The transition probabilities are learnt by the model and three behavior profiles are created: less risk, moderate risk, and high risk. If user profile matches the baseline model, it is called high model. The middle profile is partially similar to the baseline profile of monitored machine. If the user's profile matches the base model with the least probabilities, it

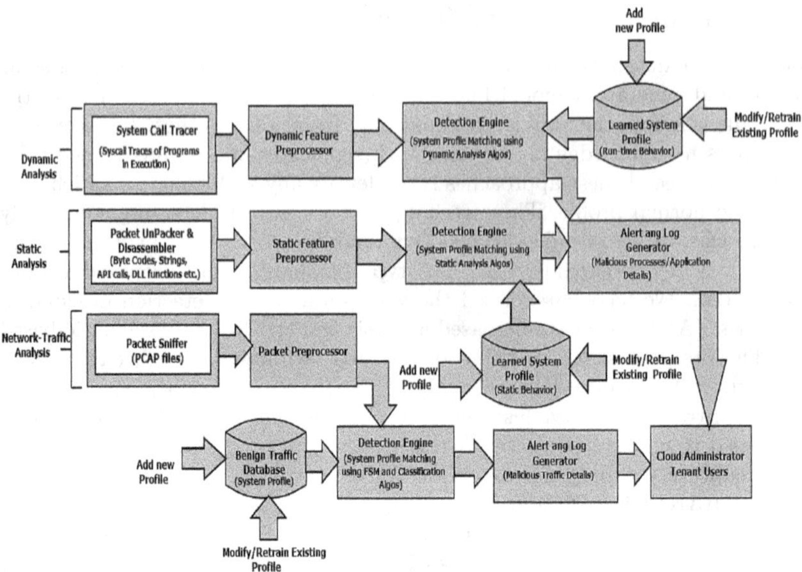

FIGURE 6.3: Conceptual working of anomaly detection approaches.

is called low profile. IDS is configured to raise the alarm whenever there is a least match of partial match with the base line model.

In ML-based models, various researchers have explored the use of unsupervised and semi-supervised ML algorithms for network anomaly detection. The prime advantage of these techniques is that they perform very well in detecting zero-day attacks (discovering vulnerabilities which have not been discovered before). The use of SVM along with hierarchical clustering is explored by some researchers [228]. The approach aims to improve the efficiency of SVM. Initially, dynamic clusters are developed using dynamically growing self-organizing tree (DGSOT). In each of the iteration, the tree grows will addition of new nodes along with training using SVM. These processes reduce the computational time. The trained model is represented by hyperplanes. The support vectors which are closed to the hyperplanes are now given as input to unsupervised algorithm. It controls the growth of the tree. Accuracy is the prime criteria used to stop the tree growth.

The eCloudIDS [190] is another intrusion detection framework. It consists of two main components of the framework: uX-Engine and sX-Engine. A log preprocessor fetches the information from log details and extracts the features which are further passed to primary engine which employs the unsupervised learning algorithm particularly self-organizing map (SOM) and learns the system's behavior. If any unexpected is observed, it talks to the permission recorder (PR) for further verification. The PR module then invokes the sX-Engine which is a decision model which have been trained using supervised machine learning algorithm. The detection engine correctly identifies the abnormal behavior. The eCloudIDS provides the 89% detection rate with a false alarm of 9%.

Various techniques are discussed to detect network attacks, etc. However, the VM-specific attacks such as cross-VM side channel, VM escape, etc., cannot be analyzed by using network traffic analysis-based approaches. These low-frequency attacks try to exploit the VM program vulnerabilities, OS vulnerabilities and hypervisor vulnerabilities, etc. Hence, by analyzing the program's behavior, one can decide whether a program is behaving suspiciously or normally. The presence and frequency of rare system calls, patterns of opcodes , system calls, API calls, etc. might be helpful for a security analyzer to get some clue about the attack behavior.

In the next section, we present the details about program behavior analysis approaches in detail:

B. Program Semantic Analysis

Researchers have employed the program semantic behavior analysis approaches for detecting malware. This type of analysis is helpful for detecting the attacks, which target the system or VM programs. Various approaches under this category have been categorized into two types: (i) Dynamic semantic analysis and (ii) Static semantic analysis.

a. Dynamic Semantic Analysis

It monitors the execution traces of the programs. The program execution traces having system call logs in various possible scenarios of normal execution are captured. These log files are pre-processed and important features are extracted. One of the key benefits of this approach is that they capture the actual behavior of monitored programs even if the program is encrypted or obfuscated. These category of approaches have been classified into four classes: (i) Enumeration-based, (ii) Frequency-based, (iii) State-based, (iv) Machine learning-based. Researchers used all four category of techniques in detecting intrusions in cloud environment. The details are given below.

(i) Enumeration-based Dynamic Semantic Analysis

These categories of approaches have been used by researchers since a very long time for detecting host based anomalies. The execution traces of the monitored programs are very helpful in identifying the system behavior.

Forrest et al. [229] first proposed the system call based pattern matching method many years ago which they have termed as look-ahead pair '' approach. This approach fetches the program features in form of two values: system call and system call sequence following, extracted using a window of n size. All unique system calls with their immediate system call sequences are recorded in the database. This database acts as a baseline profile for the future behavior of monitored programs. Any mismatch from the recorded behavior from a certain threshold raises alarm to the user.

They further extended their work [230] and proposed a new approach, called STIDE. In STIDE, short sequences are extracted from the complete long trace. In each iteration, the sliding window is shifted by 1 and next system call pattern is extracted with the same size n. If $n = 3$ then various short system call patterns having length 3 are produced by STIDE. It is observed that short system call sequences have better discriminative power than previous approach.

Gupta et al. [197] applied the above approach for detecting privilege escalation attacks. It uses a database containing the system call structures in form of key value pair (baseline database). All monitored traces of the programs are converted in form of key value pair, which are then matched again the baseline database. Any mismatch signals the suspicious behavior of the user and cloud administrators are warned for same. ISCS achieves 98% accuracy.

In their further work [172], the author proposed malicious system call detection approach called MSCS, which focus on system-wide detection. Each VM contains a program-wide VM-specific profile and cloud managing server contains the system-wide profile of all VMs. The current behavior of the VM is matched with baseline database profile created and stored in the tenant VM. It is called as program-wide detection. In case of system wide detection, the current behavior of all the VMs are matched with the respective tenant VM snapshot stored at the cloud managing server. UNM dataset [231] has been used for validation and results do not seem to be very promising.

(ii) Frequency-based Dynamic Semantic Analysis

Researchers observed that a program executes the rare patterns of the system calls while executing some malicious code or by the wrong usage of the system such as buffer overflowing, unwanted system errors, invocation of rarely used system functionalities. Hence, just by looking at the rare system. Therefore just by a mismatch of a pattern with the baseline database, we can't say whether an activity is anomalous or not. It might happen that a rare pattern is found with very low frequency count of occurrence. Hence, frequency based approaches give the importance to the count of occurrence of short system call patterns. These approaches tries to find out the frequency distribution of patterns and analyze them:

Warrender et al. [232] proposed (T-STIDE) called as time delay STIDE which adds the frequency threshold. If a mismatch, then frequency of mismatched patterns is counted. If it's lower than the certain threshold, it is then treated as a rare sequence and administrator is warned about same. The system maintains a locality frame count (LFC). All the rare sequences (mismatched) are matched against LFC. If LFC is 10; it means how many if last 10 sequences are mismatched. Kang et al. [233] employ the machine learning approaches and integrated them with frequency-based method, called "Bag of system calls (BoS)". A baseline profile is developed which represents the system call distribution in a trace in both intrusive and normal scenario. It provides good results in detecting malware.

The above-mentioned approach is adopted by Alarifi and Wolthusen [162] for malware detection in cloud environment. In the first phase, program execution logs are collected when VMs were first initialized for some duration. The have collected IOCTL system calls for some duration of time in KVM servers. A Linux command line tool named *strace* is used for this purpose. The time complexity of their approach is O(n) where no. of input lines are denoted by n. They achieved 100% detection rate with 11.11% FPR using the window size 6. The approach may not perform well in all variants of attacks as ordering have not been considered.

(iii) Finite Sate Machine-based Dynamic Semantic Analysis

Some researchers [234] proposed malware detection technique which uses Hidden Markov Model (HMM) which models the normal system behavior in form of state transition analysis of various program execution patterns. HMM-based model takes longer time in learning the behavior of all processes. Therefore, authors have considered only privileged programs to improve the efficiency of the model. When a user program invokes privilege program, the effective user id (EUID) changes to root. Once privileged operations are performed, the state of the system changes. Now the EUID becomes the normal user id. However, in case of malicious execution of the program, EUID does not change even if the privileged operations have been performed. By taking control of root access, an attacker can launch more attacks. State transition analysis based approaches model such behavior of the system. An abnormal change in the state sequence is termed as intrusion.

Yin et al. [235] employed HMM for malware analysis in cloud environment. Earlier approaches were time consuming. In addition, making the profile for all possible programs which are constantly updated, is difficult. They found that stability is found in temporal signatures and therefore such behavior is considered for achieving better efficiency. Hence, the temporal behavior of the normal program is trained by using the HMM. The major limitations with the state transition based approaches are that they consumes a lot of time in training and sometimes generates complex graphs. However, they achieve good accuracy.

Some researchers have applied the state transition analysis approaches for malware detection in cloud environment.

Alarifi and Wolthusen [182] proposed the use of HMM which captures and analyses the system call sequence of the privileged programs running inside the cloud VMs. The approach uses the Baum-Welch algorithm for making the state transition graph of various possible states of the programs. The state transition probabilities are calculated for each of the system call transition. Various different scenarios have been considered for the training. In first case, all types of system calls have been considered. In second case, mainly IOCTL system calls have been considered for creating the normal profile of the machine. They have created a self-generated dataset of DoS attack in the test bed setup of virtual environment. While considering all possible system calls, they got 100% accuracy result with 5.66% false alarms. They achieved 83% detection rate while considering only IOCTL system calls .

The above discussed techniques produce more false alarms for identification of attack variants and novel malware attacks. Hence, machine learning approaches have been integrated with the existing approaches to overcome the limitations. Various such techniques have been described below:

(iv) Machine Learning-based Dynamic Semantic Analysis

In this category of techniques, the system call patterns of the normal programs have been collected and pre-processed to extract the meaningful information. Lee et al. [236] employed the RIPPER algorithm for learning the pattern of the execution traces. The features are extracted using the sliding window concept. In each shift of the window, fixed length short sequence of system calls are extracted which are called as n-grams. The unique n-grams act as the basic building block in identifying the normal and intrusive activities. RIPPER algorithm classifies the incoming input pattern into abnormal or normal behavior. The approach also does some post-processing in which the number of input instances are counted which are classified as suspicious. If the count exceeds some predefined threshold; the region is declared as abnormal. They have considered that if the total count of abnormal regions exceeds beyond 2%, then trace is malicious.

Arshad et al. [185] organizes the system calls based on the security requirements, as discussed REMUS classification. Their attack detection system uses the misuse detection approach to learn and detect the pre-defined attacks and identify the suspicious system calls. Once a system call is classified as intrusive or suspicious, it is passed to a severity analysis system (SAS) for finding out its severity. The SAS system uses the decision tree algorithm for this purpose. The SAS module also takes the information from the VM to better understand the various security characteristics of VM. The approach achieves an accuracy of 90.79% on the self-generated dataset. The authors have not provided any details about the system call extraction from outside.

b. Static Semantic Analysis-based Approaches (SSAA)

These category of techniques analyze the static features of program without the need of executing them. The programs features are extracted inform of function names, opcodes, bytecodes and API calls, etc. The extracted features are pre-processed and important features are derived from logs. There are further two categories: (i) Specification-based and (ii) Profile-based. The details of each of the category is given below.

(i) Specification-based Static Analysis Approaches

Some researchers [237] proposed specification-based system which maintains the specification policies for each monitored program. A program intended behavior can be identified by looking at its specification. A program specification is defined in form of policies. For example, A malicious user can manipulate the remote file distribution program (rdist) which maintains the consistency of files which are stored in different systems. A normal user invokes rdist when he/she wants to do some changes in the file. The rdist program allows the modification in a temporary file before finalizing the change in original file. If attacker can fool rdist program by creating a symbolic link (name is same as of temporary file). When rdist program changes the permission (mainly at the time of final write), the permission of malicious file also changes which attacker can misuse. Now in the approach proposed by Ko et al., a specification policy is maintained for each program. Whenever a program is behaving as per specification, it means something malicious activity has happened. The major limitation with such approaches is that expertise knowledge is required to make the specification file of each monitored program.

These approaches are less suitable for cloud since it is very difficult to maintain the specification of each and each program in such complex distributed computing environment. The time to time maintenance of such systems will increase the overhead of the administrator. Moreover, it is a very time consuming task to write such policies and chances of errors are also high.

(ii) Profile-based Static Analysis Approaches

Profile-based Static Analysis Approaches These types of approaches generate a common behavior of the program by extracting and learning the static features of the program. The various such features are control flow graph (CFG), API calls, opcode sequence, and blocks of hexcodes, etc. Then machine learning is applied to learn these behavioral patterns. Below, some static analysis approaches are described.

Yuxin et al. [238] proposed two static feature extraction methods for deriving the n-gram based features from a dataset. Initially, they decompiled the executable code and generated assembly code. Once all assembly codes are extracted of each of the executable, a CFG is developed which is then converted to a running tree. A tree represents the execution flow of the code. Each of the branches of the tree is traversed to create a long sequence which are converted into various short sequences. They have used two document frequency methods: Information Gain (IG) and Document Frequency (DF). They have validated the approach with various ML algorithms. They found that DF is providing the better results with all the classifiers.

Data mining algorithms are also used in some security frameworks [239] to detect the anomalous sequences. It makes the system profile based on dynamic features such as Dynamic Link Library (DLL), function calls, and various other calls extracted by using libBFD utility. They have used RIPPER algorithm for learning the execution behavior of programs. They achieves 71.05% detection rate with 7% false alarms. They further created another dataset from non-PE executable and extracted GNU string features. Naive Bayes classifier is used for learning profile, created by using new features. They have achieved 97.43% detection rate and 3.8% false alarms.

Masud et al. [240] used the static features of monitored programs such as binary features, assembly features and DLL features, etc. The binary features are basically bycode sequences of the programs extracted using Hexdump utility. Next the assemble features in form of opcodes are extracted by using PEDisassem. This tool is also used to extracted DLL function calls. They have organized all the extracted features in form of AVL tree. They have further applied feature selection by using IG method. The constructed optimized feature vector set consists of various types of features, representing the profile of the machine. They have employed the SVM classifier.

In cloud, there exist different VMs running in different OSes. Hence, different VM will have a different set of executable codes. Hence, each of the VM specific profile need to be created and maintained by cloud administrator for applying such approached in cloud.

6.2.3 Virtual machine introspection (VMI) techniques

The above mentioned techniques are well suited to be used at VM-level or network level. However, the new generation of malware can easily detect the presence of analysis component or even the virtual environment. Hence, there are chances that the analysis tools/security tools running at VM-level may get easily evaded by advanced malicious programs. It raises the need of VM monitoring from hypervisor, a more trustworthy place. VMI techniques are used to introspect the VM and analysis the VM state information from hypervisor. Based on the detection mechanism, the approach has been categorized into five categories, explained below in detail.

(i) Guest OS Hook-based VMI

These category of approaches inject the hooks inside the VM which are kernel modules that monitor the VM states. These hooks intercept the VM events and communicate this information to the core VMI components running on hypervisor. The major limitation with these approaches is that they require the guest OS kernel modification. The VM hook modules require the special protection to avoid the unwanted access from VM applications [202].

(ii) VM State Access-based VMI

These category of techniques try to find the information related to VM states such as VM CPU registers, memory space, I/O access, network artifacts, etc. by doing the VM memory mapping from outside. The information is further analyzed for extracting high level information about VM [241].

(iii) Hypercall Authentication-based VMI

These categories of techniques intercept the software traps which are generated from kernel to hypervisor layer. Hypercalls are privileged calls generated from virtual domain to avail any hardware service in para virtualized environment. The hypercalls integrity is checked by these solutions before forwarding them to further stage [173].

(iv) Kernel Debugging-based VMI

These category of approaches are based on analyzing the kernel debug data, extracted using kernel debugging tools. The techniques extract the address of various kernel functions and inject break points before their name. The traps generated by break point signals the execution of kernel functions [242].

(v) Interrupt-based VMI

These category of approaches monitor the traps generated in the monitored VM. These techniques can be classified based on two categories: (a) Handling the Interrupts and (b) Forcing the interrupts. Let us now provide a general idea about both categories:

- *Handling the Interrupts:* Some VMI mechanisms are based on analyzing the interrupts such as VM_EXIT events, VM_Enter events, etc. and fetching the details about same from corresponding CPU registers and memory [175].
- *Forcing the interrupts:* Some VMI mechanims forcefully generate interrupts using hardware-based hooks (e.g., setting/unsetting specific register values). They further fetch the process values having VM information [243].

VMI techniques provide introspection based functionalities to analysis the VM code from outside. Researchers are working for the integration of VMI with traditional intrusion detection techniques for making them more efficient and practical. The techniques are described in detail in Chapter 8.

6.2.4 Hypervisor introspection-based techniques

There are some other introspection-based approaches called as Hypervisor introspection (HVI). HVI techniques address the attacks below virtualization layer. These techniques are primarily designed to make the hypervisor secure. If a hypervisor is compromized, all the VMs running above it will be compromized. Hence, there is a strong need to have some hypevisor security techniques in place in cloud. The HVI approaches are categorized into following types as described on the following pages:

(i) Nested Virtualization

In this kind of approaches, the security crucial portion of VMM is separated from the core functions of the VMM by incorporating the concept of nested virtualization [244]. It improves the robustness of the system and makes the system difficult to breach hypervisor-related attacks.

(ii) Code Integrity Checking Using Hardware-Support

The hardware supported schemes do not require the insertion of any additional layer. Rather they integrates the security solutions with hardware. They are of two types:

- *Snapshot based:* These techniques are based on collecting the memory and CPU snapshots. System Management Mode (SMM) is used for this purpose. However, transient attacks which targets between snapshots can affect such approaches.
- *Snooping based:* Snooping techniques overcome the limitations with snapshot based techniques. Snooper collects the memory bus traffic of the monitored machine. The assembled data is then sent to the verifier for checking the integrating of memory contents [245].

(iii) Memory Integrity Checking Using Hardware/Software Support

The dependency on hardware components like PCI, DMA, and SMM, etc. for security check makes the above approach less suitable. It's mandatory to modify the hardware architecture to support such monitoring. Hence, hardware-software based approaches are introduced to minimize the modification.

(iv) Revisiting the VMM Design

Some of the HVI solutions are based on doing amendments in the existing hypervisor architecture with an aim to reduce make it safer from attackers and reduce the need of special costly hardware [246].

(v) VM-assisted Hypervisor Introspection

Most of present techniques depend on specific feature of hardware design. Checking the integrity of control data is the main focus of such approaches. However, these techniques emphasize the need of monitoring data (non-control) like decision making and configuration data of VMM [247]. More details are given in Chapter 8. Let us now discuss hybrid techniques in detail.

6.2.5 Hybrid techniques

Hybrid approaches are introduced to improve the efficiency and effectiveness of the IDS. They combine two or more above explained intrusion detection approaches for analysis. The use of anomaly and misuse detection approaches at hypervisor level is not possible without the support of introspection based approaches. VMI techniques perform the introspection to gain insight into VM and extract meaningful information. The extracted logs are analyzed by the security analyzer to check any

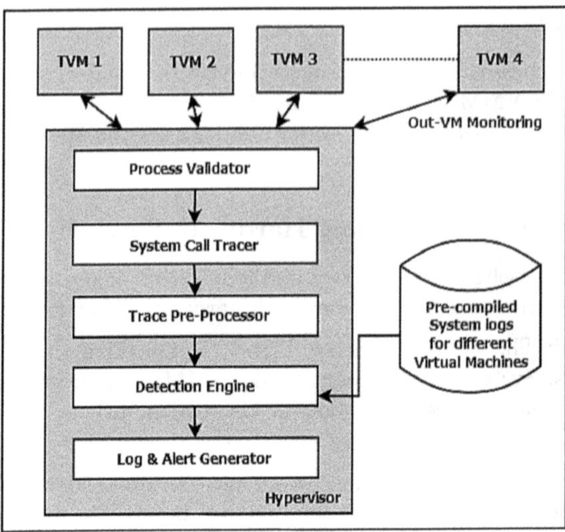

FIGURE 6.4: VMGuard introspection-based security approach (integrated with machine learning).

abnormality. The security analyzer can use any of the misuse or anomaly detection mechanism discussed in earlier sections for attack detection in cloud.

The reliability of the detection mechanism for signature based IDS is improved by Modi et al. [191]. They have integrated signature based module with anomaly detection module. They have used SNORT as signature based tool. They have used DT algorithm as a anomaly detection module, to identify the attack behavior. Initially, Snort is used to capture and process the packets and match the signatures against pre-defined database. Any network traffic passed by Snort is again analyzed by anomaly module as a second level security check. Otherwise, if Snort detects any suspicious flow, an alarm is raised to the admin. The next module matches the current behavior of the system with the learned anomalous behavior. The technique has been validated with NSL-KDD'99 dataset and achieves 84.31% accuracy with 1.91% FPR. They further improved their work and proposed a hybrid approach which integrates Snort with apriori algorithm [248]. Apriori works better than decision tree and provides good accuracy. It improves the effectiveness of the NIDS. Signature based module detect the attacks and anomaly module detects the derivatives of the attacks. Anomaly module also generates the rules and updates the signature database. They used KDD-99 based dataset for traffic validation which does not carry any virtualization characteristics.

VMGuard [168] is one of the introspection-based malware detection frameworks which makes use of misuse detection approaches and integrate them with ML and introspection. The security analyzer is placed in the privileged domain of cloud server where user VMs are hosted, as shown in Figure 6.4. It makes use of break point injection to trap the execution of monitored programs and extract the execution traces. The extracted traces are parsed and converted into numerical features. It uses Random Forest to learn and develop the intrusive profile of VM programs which is later used to decide the malicious activity in the VM.

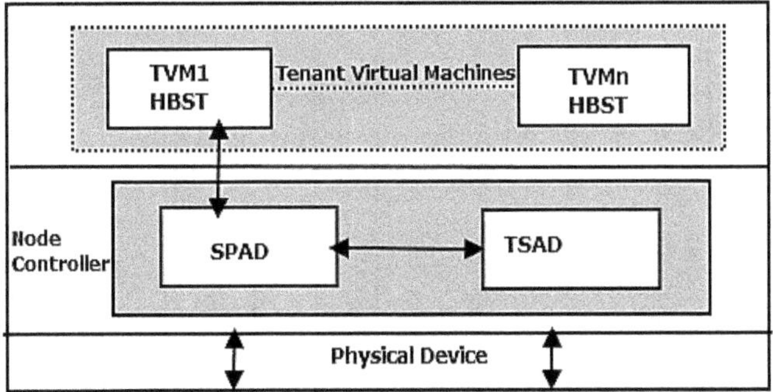

FIGURE 6.5: Basic cloud security architecture.

An attack resilience architecture is given by Watson et al. [204] for detecting attacks in cloud environment. The architecture can detect both malware and DoS network attack. The system level features such as no. of threads running in a VM, VM memory utilization, CPU utilization, etc. are obtained by making use of introspection library, i.e. LibVMI. The network-level features are extracted by launching the DoS attack and extracting the traces using networking tools. The combined features (system-level and network-level) of the attacking malware are used to train the machine learning algorithm, more specifically one-class SVM. It achieves the 90% detection rate and is applicable to detect stealthy attacks from the hypervisor domain.

Vardharajan et al. [217] proposed a security framework for cloud environment with two components SPAD and TSAD, as shown in Figure 6.5. SPAD is developed to provide baseline security. If tenants ask for additional security, CSP can apply TSAD which performs advance level security check to detect the intrusions. Tenants are charged based on the type of security opted by them. The network traffic is validated for the presence of any spoofed traffic by SPAD. It also does the secure recording of network traffic. The recorded logs can be helpful in resolving the billing conflicts between tenants and CSP. The XenStore VMI library is used at the time of process validation. The model has been validated for insider and denial of service attack. The technique takes the assumption that VMM is trusted and is limited to detecting VM-level attacks. The architecture provides an overhead of 6%.

Although the integration of misuse detection with anomaly detection improves the efficiency of detection algorithm. The use of VMI with these approaches makes the detection algorithm more powerful. Introspection-based approaches provide the VM state information and other details to the analyzer for further analysis. These techniques are applicable to be applied for host audit log analysis that represents the behavior profile of user. The summary is shown in Table 6.1.

TABLE 6.1: Summary of Cloud-IDS

Author	Attack Detection	Approach	Placement	Key Algo	Limitations
[187]	Network attacks (VM)	Signature Matching	VM	Snort	-Can't detect zero day attack -Regular maintenance of signature database -More False alarms
[246]	Side Channel Attacks (VM)	Hardware-assisted virtualization (HVI)	No VMM (design strategy)	Hardware-aware device virtualization	- May prone to DMA and SMM rootkit attacks - Limited in Performance
[215]	Data Leakage (VMM/VM)	Nested Virtualization (HVI)	Below Hypervisor	MD5 hash and Merkle tree	- Can't detect hardware attacks - Maintaining two EPTs page impose overhead
[191]	Network Attacks (VM)	Signature Matching	VM, Cloud controller, Cluster controller	Snort	-Can't detect zero day attack -Regular maintenance of signature database -More False alarms
[192]	Network Attacks (VM/VMM) Hidden Rootkits	Signature Matching and Profile based approach With VMI functions	VMM	Snort, Xen Access library	-May prone to kernel manipulation attacks -Regular maintenance of signature database
[190]	Network Attacks (VM)	Machine Learning approach	VM	SOM	- An attacker can subvert an IDS - More False positives
[209]	Network Attacks (VM)	Machine Learning approach	VM	ML Algo (ANN)	-More false alarms -Vulnerable to privilege escalation attacks
[191]	Network Attacks (VM)	Signature Matching and anomaly detection	VM, Cloud controller, Cluster controller	Snort and ML Algo	-More false negative -Can't detect VM specific attack -Regular maintenance of signature database
[162]	Malware Attacks (VM)	Frequency based approach	VM	Bag of System calls	More false alarms -Less accurate for Network attacks
[198]	Network Attacks (VM/VMM)	Signature Matching and VMI	VMM	Snort	-Can't detect zero day attacks -Regular maintenance of signature database -More False alarms
[202]	Malware Attacks (rootkits) (VM/VMM)	Binary signature analysis, VMI	VM (guest VM and privilege VM)	Python daemon and security analysis tool	-overhead in keeping track of register updates -Vulnerable to kernel attacks.

Continued.

TABLE 6.1: Continued.

Author	Attack Detection	Approach	Placement	Key Algo	Limitations
[216]	Hardware attacks (low-level attacks)	Memory Integrity checking	VMM	AISE and BMT	-Can't prevent against side channel attacks. -Requires modification in MMU
[185]	Malware and Network attacks (VM/VMM)	Machine Learning approach	VMM	Decision Tree (C 4.5)	- Rigid mapping of system call to specific attack may increase false positives. - Lack in providing technical details of system call tracing at VMM.
[199]	Network Attacks (VM)	Signature Matching, Profile based	VM	Snort and DDSGA analyzer Algo	-Maintenance -More False alarms -can't detect VM specific attack
[171]	Network Attacks (VM/Cloud Servers)	Pattern Matching and anomaly detection	VM, Cloud controller, Cluster controller	Snort and ML Algo (DT, Apriori)	-More false negative -can't detect VM particular attack -Regular maintenance of signature database
[182]	Malware Attacks' (VM)	Frequency based approach	VM	Bag of System calls and HMM Model	-Less Accuracy -More False alarms -Vulnerable to kernel attacks
[217]	Network and Malware Attacks (VM/VMM)	Pattern Matching and Profile based approach With VMI functions	VMM	Snort, Xen Access library	-Vulnerable to hypervisor rootkits. -Regular maintenance of signature database
[197]	Malware Attacks (VM)	Enumeration approach	VM	ISCS	-Huge storage -False alarms -Vulnerable to kernel attacks
[193]	Network Attacks (VM/VMM)	Machine Learning approach	VMM	FCN-ANN Algo	- May prone to kernel manipulation attacks -More false alarms
[249]	Hypercall-based attacks (VM/VMM)	Nested Virtualization	Below Hypervisor	Break point Injection and stide	- Can't detect system call attacks -Additional VMM layer increases Trusted Computing Base (TCB)and hence vulnerability surface is also increased

6.3 Conclusion

In this chapter, various intrusion detection techniques have been discussed in detail. The intrusion detection techniques are classified into five different categories. In each category, various techniques are discussed with various examples of security frameworks. Misuse detection and anomaly detection are the traditional IDS approaches which are applied by researchers in cloud environment. However, the traditional approaches may not be capable enough to detect the VM-specific attacks in cloud such as VM Escape attack, side-channel attacks, hyperjacking attack and VMM rootkits, etc. Hence, researchers applied introspection based approaches to detect the virtualization specific attacks. These frameworks are hypervisor dependent and carry certain overhead which restrict their use in all types of security requirements. Hybrid approaches use two or more IDS categories to generate a hybrid model and proved to be more efficient.

6.4 Questions

Fill in the blanks

1. Mark the correct statement. The hypervisor-based security analyzer is

   i CloudVisor
   ii ISCS
   iii BoS
   iv None of these

2. The signature matching approaches perform well in detecting

   i Known attacks
   ii Zero-day attacks
   iii Unknown attacks
   iv None of these

3. The run time behavior of the programs is observed in

   i dynamic analysis
   ii static analysis
   iii specification analysis
   iv both i & ii

4. Mark the approaches to detect malware attack

   i Bag of System calls (BoS)
   ii Immediate system call sequence detection
   iii both i & ii
   iv None of the above

5. Mark the incorrect statement

   i In misuse detection techniques, an attack database of prior known attack is maintained.
   ii Network security based approaches capture the ingress and out-gress traffic from monitored network.
   iii In dynamic analysis approaches, the program behavior is observed without running them.
   iv VMI techniques are used to introspect the VM from hypervisor.

Short-Answer Questions

1. Explain the misuse detection techniques and their usage in cloud security.
2. Define 'sematic gap problem'. Compare virtual machine introspection and hypervisor introspection.
3. How Could-IDS have been evolved over time?

Long-Answer Questions

1. How traditional IDS techniques are different from cloud-specific IDS techniques.
2. Define intrusion detection techniques. Explain anomaly detection techniques with suitable example.

Part III

Tools and Advances

Chapter 7

Overview of Tools (Attack/Security) in Cloud

7.1 Introduction

Cloud security [250] is the group of the processes, standards, and procedures that are designed to provide security in cloud. It manages both the logical and physical-level security issues across all platforms and infrastructures. The key sub-domains of cloud security are network-level security, Virtual Machine (VM)-level security and Virtual Machine Monitor (VMM)-level security. Network security [251] is very important for both the customers and cloud provider's perspective for providing a secure network over which the data can be securely transferred from one end to another. The network administrator is responsible for maintaining their network security by adopting different preventive measures to protect their network from various security threats. The security of the network is maintained at three levels—physical, technical and administrator level. Network attacks are launched at this level and hence proper network security tools should be in place.

VMs are prone to be attacked by malicious tenant users. VM-level security ensures the safety, privacy and availability to the tenant's data stored in the cloud environment. At this level, attackers use the cloud to launch the malicious applications that perform attacks against other tenants of the cloud, or the cloud itself. Thus protecting the integrity and confidentiality of the data on cloud becomes a major concern. VM-level attacks are tackled by the VM-level security tools. At VMM-level, attacks exploit the vulnerability of the hypervisor. For example, an attacker can install a malicious/rough hypervisor to take the control of the server and manipulate it. This attack is called hyperjacking attack. Such attacks are caused by rootkits-based malicious software. They exploit the virtualization features and take over higher privileges than the domain and host of the target. Attacks on the hypervisor are also performed by taking the backdoor entry into the hypervisor's domain and overwriting the integral code and thus taking control over the kernel data of the OS. Thus, VMM-level security tool should be deployed to ensure hypervisor security.

In this chapter, a deep insight is provided on various types of attack and security tools in cloud environment. A threat model has also been discussed which describes various attack surface from where attack can be launched. We further provide a detailed classification of attack tools based on the vulnerability exploited by them at VM-level, VMM-level, and network-level. Network-level attacks are performed by powerful tools like XOIC [252], RUDY [253], DDosSIM [254], etc., that cause disruption of the web server and thieving private data or information. VM-level

DOI: 10.1201/9781003004486-7

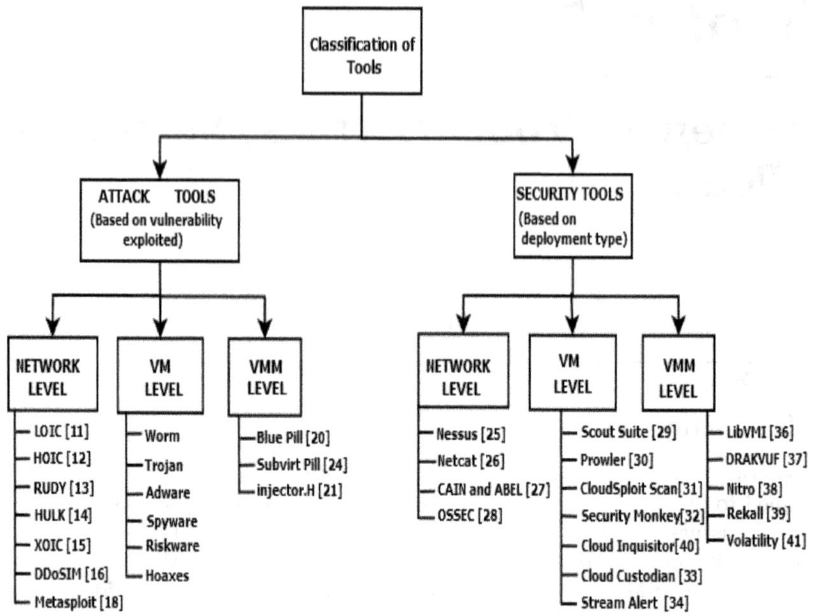

FIGURE 7.1: Classification of tools.

attacks are host based attacks that target the root of the system. Some tools like worms, trojans, viruses, etc., penetrate the application layer to cause malicious activities. Attack tools like Blue Pill and Subvirt Pill [255] and Hinjector [256] attack at the VMM level. These tools perform hyper-jacking by attacking the hypervisor and cause a disruption at the OS level. To detect such attacks, there is a dire need of having appropriate security tool in place at each level (i.e., VM-level, VMM level, and Network level) in cloud environment. Based on the deployment type, we provide a detailed classification of security tools that are configured for increasing the safety against attacks at each level in cloud. These security tools provide a wall of security from the attacks, performed by hackers and intruders.

A brief classification of both the attack and security tools is shown in Figure 7.1. Networking tools such as Nessus [257], Netcat [258], OSSEC [259] ensure the safety from network attacks. Security at VM-level is maintained by tools such as Prowler [260], Security Monkey [261], Cloud Inquisitor [262], which prevent the attacks from disrupting the applications at host level, while tools like LibVMI [164], Drakvuf [166], nitro [224] prevent breaches at the VMM-level. All the attack tools do attack systems in different ways. Each security tool has specific procedure for detecting specific patterns of threats and attacks. There is no such tool that can detect all variety of attacks. Such variety of tools can be helpful to the researchers working in the area of cloud security. Research can create their own dataset by having knowledge of such tools. It also helps researchers in developing new and advanced tools.

7.2 Attack Tools

Attack tools consist of code and rules that were made to disrupt network and computer systems. Attacks on servers and systems are made to get unauthorized access to private data or to exploit the resources of the target. There are various attack tools that are used to cause disruption to the target server or system and they have been summarized in Table 7.1 with all the necessary requirement and information.

7.2.1 Network-level attack tools

Networks are liable to be targeted from malicious sources. An attack on network can happen when an intruder deviates data traveling through a network. An intruder can also initiate commands to cause disruption of normal operations or activities of the network to conduct lateral movement in order to find and gain access to assets of the users. The following are some tools that are used to attack and hack the network's policy and practices for prevention of unauthorized activities on the network.

Low Orbit ION Cannon

It is a free open source network stressing tool [263] which is used to attack the network easily. In this attack, the server is flooded with HTTP, TCP, and UDP packets. The heavy web traffic causes disruption of services of the target website. This tool provides a simple GUI, where the user only needs to input the target address of the website and type of flood request to be generated. The limitation of this tool is that the attackers cannot use proxy servers to generate packets. Therefore, it is difficult to obfuscate the IP address of the attackers. Thus, it is easily traceable and it is not suitable to perform Denial of Service attacks. Also it cannot attack more than one user concomitantly.

High Orbit ION Cannon

It is an open source, stress testing tool [264] to attack the victim's network by flooding it with junk GET and POST request packets causing an HTTP flood attack. It is the successor of a DDos attack tool, LOIC, and unlike LOIC it can attack multiple targets concurrently. The attack technique of this tool is same as LOIC which is used to overload the victim's server and cause it to shutdown its services. This tool provides a simple, user friendly graphic interface thus it is easy to be used by users having less knowledge. Due to this feature, it also becomes dangerous tool as misuse of the utilities to disrupt services of website becomes facile. Though, Add-on scripts are used to hide the location of the attacker in HOIC, the traffic generated by a single attacker is still not enough to disrupt the services of the target network. However like LOIC, it cannot generate TCP and UDP floods.

TABLE 7.1: Comparative analysis of attack tools

Attack Tools	Release	Environment	Interface	Language	Category
LOIC[263]	2008	Windows, Linux, OS X, Android, iOS	GUI	C#, C++	Network
HOIC[264]	2012	Windows, Linux, OS X	GUI	Visual Basic, C#	Network
RUDY[253]	NA	Windows, Linux, macOS	GUI	Python	Network
HULK[265]	2012	Windows, Linux	GUI & CLI	Python	Network
XOIC[252]	2014	Windows	GUI	Python	Network
DDoSIM[254]	NA	Linux	GUI	C++	Network
PyLoris[266]	2009	Linux, Windows, OS X	GUI	Python, shell script	Network
Metasploit[267]	2007	Linux, Windows, OS X	CLI	Ruby	Network
dsniff[268]	2000	Unix – like	CLI	NA	Network
Blue Pill [255]	2006	Windows	CLI	NA	VMM
Subvert Pill	NA	Linux	CLI	NA	VMM
injector.H[256]	2013	Windows, Linux	CLI	NA	VMM

RUDY

The term 'RUDY' [253] refers to 'R U Dead Yet is a DDos' attack tool. It executes low and slow rate attacks. It crashes the server by submitting long form fields unlike disrupting the network server with a lot of packets requests. When these forms are detected, a HTTP POST request is sent with an extensive 'content-length' header field to avoid the server to close the connection. It starts injecting the form with data, one byte packet at a time at a very slow rate with larger time gaps between successive bytes. At the server end, these huge amount of packets use all the resources causing a shortage. Thus it causes the application not to meet any legitimate request. This tool only works at the HTTP based layer (7th layer) and also the attack happens at such a slow rate that consumes a large amount of time. Data with such slow rate can be detected as abnormal traffic which can be blocked easily.

HULK

The term 'HULK' [265] refers to 'HTTP Unbearable Load King' which is a distributed denial of service tool that generates large numbers of unique HTTP GET requests. The network requests are distinguished by having an arbitrary value for URL parameter and header for each request. This tool uses such obfuscating technique to sidestep the caching engines (which can detect and block abnormal traffic) and attack the web server's resource pool. This tool works well in case of existence of behavioral analysis in the network. Along with obfuscation of traffic, this tools also hides the source address of attack and capable of forging web requests for various fields. However, the request traffic from this tool can be studied and proper ground rules can be laid to block the attacks.

XOIC

XOIC [252] is a DDos tool which is used to attack the web server. It is more powerful than LOIC. It is easy to use, even for the beginners as it has a simple and user-friendly interface. XOIC has three attack modes. The first mode is used to request the IP address, port number, and protocol selection fields. The next mode is used to launch the normal denial of service attack and the last mode is used to launch various DoS attacks with ICMP TCP UDP HTTP messages. The limitations for this tool is that the requests can easily be detected and blocked since there are no additional features available to obfuscate the traffic requests.

DDoSIM

DDoS Simulation attacks [254] are executed using this tool on various networks and websites. This tool can perform application layer attacks, volumetric attacks and even resource starvation DOS attacks. During the application layer denial of service attack, the layer 7 applications of the OSL model are affected. This attack can be done by generating Mail, SIP, or DNS query floods to disrupt the whole application or the server. In case of volumetric attacks, the end-targets are affected as floods of UDP packets, SYN and ICMP messages attack the server. The resource starvation DoS attacks are also called Low-and-Slow attacks as they attack the network stack or OS resources through SMTP mail requests, NTP or DNS requests in huge number.

PyLoris

This tool [266] is written in python and is used for testing the vulnerabilities for a DoS attack. It makes the simple - open full TCP connections and keep them open for a long duration of time which causes DoS to the legitimate requests. PyLoris also avails SOCKS proxies and SSL protocol connections. It can also target the protocols like FTP, IMAP, HTTP, SMTP, and Telnet. The flaw of PyLoris is that it is a scripted tool and supports limited functionalities.

Metasploit

The Metasploit [267] is an open source network security framework that provides a free public resource to research and gather information about security vulnerabilities. It generates a code for the web server administrator to breach into his own network. The administrator further identifies and document the security risks which are on top priorities with respect to the security of the network. This framework aids in penetration (pen) testing of software also in signature development for security tools. It also provides tools and resources for automatizing the process of comparing the network administrator's code risks of repaired versions. This framework also includes anti-forensic and advanced evasion tools. Metasploit comes pre-installed in the Kali Linux OS.

dSniff

Dsniff [268] is a collection of tools which are used for password sniffing and network traffic analysis. It was written by a network security researcher to analyze protocols for different applications and to extract necessary information. Tools like filesnarf, msgsnarf, webspy, mailsnarf, dsniff, and urlsnarf keep a check for stimulated data over the network like e-mails, passwords, files, etc. Whereas tools such as macof, dnsspoof, and arpspoof aid to the interference of the web server traffic that is usually not available to an attacker. Tools like webmitm and sshmitm execute the man-in-the-middle attacks against the SSH and HTTPS sessions by utilizing the weak bindings in ad-hoc Public Key Infrastructure (PKI). It can also be used to exploit the standard behavior of switched networks.

7.2.2 VM-level attack tools

VM-level attack tools are designed in a manner to attack the applications in VM level or Host level. There are many such tools which attacks or penetrates the application and causes some malicious activities to take place.

Worm

A worm is a program that is independent and creates its multiple copies itself across the network. For example, a virus spreads only when an infected file is copied which clones itself into multiple files. A worm spreads strenuously by sending copies of itself over Internet, through e-mails or OS security bugs. Apart from it, worms have the potential to bring with them some extra malware. Although this kind of functionality is not strictly limited to worms. Worms are also used to obstruct and completely block the communication tracks by using DoS attacks. It has the capability to spread across the world with the help of internet within a minute. There are numerous number of worms. Some of them are—Badtrans, Bagle, klez, Hybris, Swen, Upering, etc. The tool EternalRocks [269] is used for spreading worm in a network model.

Trojan

Trojan is a malware but it cannot replicate itself and infect other files like viruses and worms. It mostly exists with .exe extension (that of an executable file) and only contains trojan code and nothing apart from it. Therefore, the only way to be safe from trojans is to delete them as soon as possible. It has variety of functions, from key logging, to delete the files or formatting the disc. Some trojans come with a unique feature which installs a back door for the attacker. For example, a client-server application can grant the developer, a remote access to the computer. Such applications can be installed in client system without the consent of the client. Various types of trojan exists in internet as well as in the system like—Backdoor, Exploit, GameTheif, Ransom, etc. There are tools which are used for creation of trojans like TheFatRat [270] which is used to create the backdoor for bypassing the browser.

Adware

The term Adware [271] stands for advertising-supported software which is solely dedicated for displaying appropriate as well as un-appropriate advertisements. When the user does surfing or browsing something on internet, Adware will display pop-up boxes and initialize some websites as the default homepage. It sometimes opens a special interface window. Adware is merged with programs that are available for downloading freely. The client computer is aware of this in the End User License Agreement which is generally ignored by the users. Adware's advertisements are the main source for the freeware developers to earn income by offering various extra things which are generally available with paid version of the software. Installing adware generally falls within the fences of legal guidelines. There exist many legit advertisement supporting programs. However, issues of the assertiveness of advertisements along with its contents make the existence of adware questionable. There are various types of adware such as—1ClickDownloader, 4-you.net Search, BasicSeek, GreyGray, Lyrmix, etc.

Spyware

The term 'Spyware' [272] is a software which uses the Internet for collecting and containing various detailed information of sensitive documents without the consent or awareness of the user. There are variety of spywares whose functions are totally different from each other. Some Spyware programs search for information such as lately installed application software. Sometimes, they are designed in such a way to collect the information of the user's browsing history. Another type of spyware programs is designed with a very dangerous aim to collect the financial or personal data for the purpose of Identity Theft.

RiskWare

It is an admissible software [273] that can cause disruption if used by malicious users. These malwares can modify, block or copy data and exploit the performance of computers and networks using this tool. Thus Riskware is also considered as a type of malware which incorporates all applications that increase the risk of security to the user when running. Riskware installation is confirmed by license agreement, like adware and spyware. Common Riskware examples are "Dialers". Dialers are programs that channel the connection to an already set paid number. Internet service payments use such programs legitimately, but they are often exploited as the averting occurs without the user's awareness.

Hoaxes

Hoaxes [274] are programs used to send wrong information intentionally through email and spread with the help of an unsuspicious or uninformed public. Malicious hoaxes mostly instruct users to delete valid and important operating system files, professing the file to be a harmful virus. Often hoaxes cite names of credible institutions or companies like "Microsoft" or "CNN", to draw attention of the user. These messages mostly apprise of ruinous or appalling consequences. These warning urge the users to send messages of these warnings to everyone which perpetuates the hoax's life-cycle. To provide security against hoaxes, one should substantiate the email messages' claims' authenticity before making any move corresponding to the message.

7.2.3 VMM attack tools

There are some attack tools which are specifically designed to attack the hypervisor or perform hyperjacking that causes the problem at the hypervisor-level. Once a hypervisor/VMM is breached, all VMs running over it can be easily compromized.

Blue Pill

The main concept of Blue Pill [255] is to catch the instance of an OS by executing a hypervisor(thin) and virtualizing the remaining systems under the host OS. The OS which was in execution phase previously, will maintain all its current references to almost everything like hardware interrupts, data requests. The most importantly time of the system can be hacked, resulting into generating fake response from the hypervisor.

Subvirt Pill

It is also known as root kit of proof-of-concept [255]. It is most vulnerable to lacunae in the security. It drops a VMM-layer below the Operating System installation. Once a targeted Operating System, is jacked up inside a Virtual Machine, it becomes very hard to detect its presence. Its state is impossible to be detected by software specially designed for security purposes running in the targeted operating system.

HInjector

There are various ways to penetrate VMM via malicious device drivers, VM exist events or malicious hypercalls. The hypercalls are nothing but traps from para VM to VMM. The hypercall drivers are installed inside the VM. A malicious guest kernel can generate the malicious traps by injecting some malicious hypercalls during the execution of VM. If any vulnerability exists in the hypercall handler, there is a high potential to generate hypercall attacks either by injecting some malicious calls or by manipulating the parameters of the hypercalls. Attacker tries to take control over the VMM memory and escalate the VMM privileges. HInjector [275, 276] is such attack framework used to inject hypercall attacks in Xen-based para virtualized environment. It is not possible to detect the hypercall attacks by traditional IDSes. There is a strong need to design security tools to handle such threats in virtualization environment.

7.3 Security Tools

One has to be very careful of the applications or scripts which causes different types of attack in VM-level or networking-level. There are various tools specially targeted for security at various levels. Following are some security tools which are used to safeguard the system and are summarized in Table 7.2.

TABLE 7.2: Comparative analysis of security tools

Security Tools	Release	Environment	Interface	Language	Category
Nessus[257]	1998	Linux, Windows, OSX	GUI	NASL (Nessus Attack Scripting Language)	Network
Netcat[258]	1995	Linux, Windows	GUI	Scripting Language	Network
Cain and Abel[277]	2003	Windows	GUI	NA	Network
Ossec[259]	2018	Linux, Windows, OSX, Solaris, FreeBSD	GUI	NA	Network
ScoutSuite[278]	2013	Cloud	CLI	Python	VM
Prowler[260]	2011	Cloud	CLI	Shell script	VM
CloudSploit[279]	2015	Cloud	CLI	NodeJS	VM
Security Monkey[261]	2014	Cloud	GUI & CLI	CPython2.7	VM
Cloud Inquisitor[280]	2017	Cloud	CLI	Python	VM
Cloud Custodian[280]	2019	Cloud	CLI	Make/C toolchain, Python	VM
StreamAlert[281]	2017	Cloud	CLI	Python	VM
Hammer[282]	2013	Cloud	CLI	Python	VM
LibVMI[164]	2013	Linux, Windows	CLI	Shell script, Python	VMM
Drakvuf[166]	2008	Windows	CLI	Shell script, Python	VMM
Nitro[224]	2013	Linux, Windows	CLI	Python	VMM
Rekall[283]	2013	Linux, Windows, OS X	CLI	Python	VMM

7.3.1 Network security tools

Network security is to provide protection to the integrity and privacy of the network and data. Effectual network security manages to access the network, identify and target various threats and prevents them from spreading into the network.

Nessus

Nessus [257] is an vulnerability scanner that uses exposures architecture and the common vulnerabilities for cross-linking easily between the compliant security tools. It uses Nessus Attack Scripting Language (NASL), a language that describes the individual potential attacks and the threats. It contains remote clients which allow the interaction with the Administrator, and the modular architecture that consists of centralized servers that are used for scanning purpose. Administrators can also use the NASL details of the suspected vulnerabilities for developing the customized scans. It is very well compatible with the various computers and the servers of various sizes. It also helps in the detection of the security holes in the local or the remote hosts, and thus in timely manner, the proper security audits are scheduled.

Netcat

It is the networking utility [258] for providing the function of reading and writing to the network connections with the help of TCP or UDP. This command is designed to be dependable back-end that could be easily driven or used directly by any other program or script. It is also a tool for providing investigation and the network debugging. It can provide almost every kind of connection which any user could require and has many in-built capabilities. It contains various features such as transferring files, port scanning and port listening, etc. It can also be used as a backdoor.

CAIN and ABEL

Cain and Abel is a password recovery tool [277] for the Microsoft Operating Systems. The passwords can be recovered by easily breaking the encrypted passwords using any dictionary, by sniffing the network or by applying the Brute-Force method for analyzing the routing protocols and recording of the VoIP conversation. It does not destroy any kind of software vulnerability or any bug that could not be resolved with a little effort. The tool also covers some security aspects or the weakness in the authentication methods, caching mechanisms, and the protocol's standards. The main focus is the simplified recovery of any password or the credentials from different sources. However, it also provides some of the "non-standard" utilities for the users of Microsoft Windows.

OSSEC

Open Source HIDS SECurity (OSSEC) [259] is an open-source, free host based intrusion detection system (HIDS). It provides the facility for the integrity checking, log analysis, root kit detection, time-based alerting, Windows registry monitoring and also the active response. It easily provides the detection for any kind of intrusion for most operating systems like the OpenBSD, FreeBSD, OS X, Solaris, Linux and the Windows. OSSEC has the centralized and the cross-platform architecture which allows multiple systems to be monitored easily and managed properly. Analysis engine of OSSEC is capable of correlating and analyzing the logs from various formats and devices. OSSEC is also compliant with the Payment Card Industry Security Standard (PCSDSS) requirements.

7.3.2 VM security tool

VM security mainly focuses on the security of applications and data stored in VM. The VM-layer is the most weakest layer of cloud because of having the direct access with the users. Malicious users can install malware with an intention to gain access the root privileges of guest machine and launch furthers attacks in cloud. Below, we describe some of the security tools:

Scout Suite

Scout Suite [278] is a multidimensional tool designed for cloud security. Cloud providers have provided various APIs for manual inspection. It gathers the configuration data and highlights threat areas. Rather than pouring through dozens of pages on the web consoles, it provides a clear view of the attack surface automatically. Though it is stable and maintained in regular interval, but a number of features and internals may change.

Prowler

Prowler [260] is a command line tool for AWS security best practices assessment, auditing, hardening and forensics Readiness tool. It follows guidelines of the CIS Amazon Web Services Foundations Benchmark (49 checks) and has 40 additional checks related to GDPR and HIPAA. Prowler helps on executing the specific checks without the execution of entire report. It can also check multiple AWS accounts in parallel.

CloudSploit Scans

To allow the detection of risks that are security related in an AWS account, an open source project known as CloudSploit [279] scans is designed, specifically to detect such type of risks. These scripts are designed to run against an AWS account and return a series of potential misconfiguration and security risks. It provides various free and paid versions of security scanner and have got many additional features.

Security Monkey

Security Monkey [261] is an open source application which monitors Amazon Web Services and Google Cloud Platform accounts for policy changes. On detection of some misconfiguration, it alerts for insecure configurations. The support is available for OpenStack public and private clouds. It watches and monitors GitHub activities like organizations, teams, and repositories. It provides a single UI to search and browser through all of the accounts, their regions, and cloud services. It remembers the previous states and shows exactly what has been changed and when. It can be extended with custom account types, custom watchers, custom auditors, and custom alerts. It works on CPython 2.7 and supports LINUX and MAC.

Cloud Inquisitor

Cloud Inquisitor [262] is used to improve the security instance of an Amazon Web Service activities by observing the AWS objects for ownership classification, notifying account owners of unowned objects, and constantly removing AWS objects which are not owned by anyone. If ownership is not yet resolved, it also detects domain hijacking and then verifies security services such as Cloudtrail and VPC Flowlogs and the IAM policies which are managed across various accounts.

Cloud Custodian

Cloud Custodian [280] is an engine of rules that manages cloud accounts which are public. It also manages their resources. Using this engine, user can define their own scripted policies that help to enable a cloud architecture which will be well managed. It is reliable and does not incur in larger costs. It consolidates many ad hoc scripts of organizations into a lightweight and flexible tool with unified metrics and reporting features. It can be used to manage Amazon Web Services, Google Cloud Platform and Azure environments by safeguarding dynamic concurrence to the security policies. For example, it manages the access requirement and encryption policies, tagging policies and manages the cost by doing garbage collection of the resources which are not used for certain time. It also manages those resources which are not used in off-hours. For providing real time enforcement of policies with built-in provisioning, it helps the provider in provisioning the server-less capabilities.

StreamAlert

StreamAlert [281] is a dynamic data framework used for security analysis. It is serverless that gives the user the power to analyze, ingest and generate alert in any environment by utilizing sources of data. Users and various programmers use StreamAlert to scan the log data which is in TB's for detecting the anomalies and responses. StreamAlert have various features. It supports python libraries as the rules are written with the help of python language. In addition, logs which are ingested and alerts which were generated earlier can be recursively searched later for further research. The serverless designs are very easy to maintain and its scalability is very high (it scales up to 1TB per day).

Hammer

Dow Jones Hammer [282] is a tool designed for securing AWS with multiple Amazon Web Services account. It is best suitable for identifying abnormal configurations and data which are not secure within the popularity of Amazon Web Services resources, across all the availability zones and AWS credentials of accounts. It has the capabilities of dynamic reporting, e.g. Slack, Jira to provide the result in the form of feedback to the programmers so that necessary steps can be taken to correct the abnormal or wrong configurations. This helps to protect products deployed on cloud by creating secure guardrails. It creates the firm guardrails for protecting the products deployed in cloud environments.

7.3.3 VMM security tools

VMM security is as important as the security of other layers in cloud. If VMM layer is compromised, an attacker can bypass the security of any VM running above it. Hence, there is a strong need to secure VMM-layer as well. Following some of the VMM-layer security tools.

LibVMI

LibVMI [164] is a virtual machine introspection library that is used for accessing the memory of any running VM. The primitives are also provided by this library for accessing the memory using any virtual or physical addresses along with the kernel symbols. The physical memory snapshot is helpful for debugging or forensic purposes. Apart from this, LibVMI also supports memory events. These events also provide the notifications when the registered regions of memory are executed, or read, or written. These type of events need the support of hypervisor and are only available with Xen. The library only supports Linux with Xen or KVM and Mac OS X (only file access). All the platforms are well tested and worth exploring. It also provides the physical memory access to virtual memory and the kernel symbols of Linux and Windows VMs. However, a huge drawback is that, it does not have significant community support with KVM.

DRAKVUF

It is a virtualization-based agentless black-box binary analysis system [166]. It allows in-depth execution tracing of operating systems, without installing some special tool or software inside virtual machine, used for security analysis. It provides very strong malware analysis as its footprints are not easily traceable from the malicious activities. Although it is not limited to malware analysis, it provides plugins available for Windows to monitor several system aspects such as trace the heap allocations, file being accessed, extracting files from memory before they are deleted and tracing UDP and TCP connections.

Nitro

It is the introspection tool [224] for KVM only. It will receive the events generated by KVM and display them. The back end is supposed to analyze raw nitro events, and extract useful information, such as process name, process PID and system call name. It is broken into 3 components, i.e., Kernel Modules, QEMU, and Nitro/libnitro. Apparently it is limited to system call trapping but in future its use can be extended to get more information out of the VMM.

Rekall

Rekall [283] is an advanced forensic and response framework. Initially, it was a purely a memory forensic framework, it has now evolved into more sophisticated platform. It implements the most advanced analysis techniques in the security field. It is developed with a free and open source license.

Volatility

The first version of the Volatility framework [284] was released in 2007, at Blavk Hat DC publicly. This software was basically based on the years spent on the published academic research in the field of advanced memory analysis and forensics. Up to that point, the digital investigations had their focus on finding the contraband within the hard drive images. Volatility brought the power of analyzing the runtime state of the system with the help of the data found in the volatile storage (RAM). It provides a modular, cross-platform and an extensible platform for the encouragement of further works in the field of research. This software extracts and inspects the memory artifacts of both the 32-bit and the 64-bit systems. Volatility also has support for all the versions and flavours of Windows, Linux, MacOS and also Android. It has the power for analyzing the raw memory dumps, virtual machine snapshots and the crash dumps. VMWare dumps (.vmem), hibernation files, virtual box dumps, Microsoft crash dumps and many more. The system can also be investigated independently by this framework and contains more than 35 plugins for analyzing. It is pre-installed in various Linux flavors/versions such as REMnux, Kali Linux, and many others. Apart from being freely available at Git, another goal is to increase the innovation, collaboration, and the accessibility for the detailed knowledge between the various offensive software communities.

```
hpc-client@hpcclient-H61MLV3:~$ egrep -c '(vmx|svm)' /proc/cpuinfo
4
hpc-client@hpcclient-H61MLV3:~$ kvm-ok
INFO: /dev/kvm exists
KVM acceleration can be used
```

FIGURE 7.2: Checking the Virtualization Support.

7.4 Case Study of LibVMI: A Virtualziation-Specific Tool

The following system configuration is used—Machine with Ubuntu 16.04, Model name: Intel(R) Core(TM) i5-3470 CPU @ 3.20GHz Architecture: x86_64, RAM: 16 GB, Guest VM running Windows 8.1, Intel-VT Support: Yes
The various steps to be followed are listed below:

7.4.1 Check the system configurations

Step 1: Upgrade you OS by running command as shown in Figure 7.2. If output is 0, it means the system doesn't support virtualization (enable it from BIOS setting if such support option (Intel-VTX) available). If output is any number greater than 0 then you are good to proceed further.

7.4.2 Install KVM and necessary dependencies

Step 2: Install KVM if Step 1 is passed along with other dependencies as shown in Figure 7.3.

```
hpc-client@hpcclient-H61MLV3:~$ sudo apt install qemu qemu-kvm libvirt-bin bridg
e-utils virt-manager
[sudo] password for hpc-client:
Reading package lists... Done
Building dependency tree
Reading state information... Done
```

FIGURE 7.3: KVM Installation.

Step 3: Check the system architecture by running following command: '*arch*' In our case, output is x86_64 which represents 64 bit kernel machine. The architectures: i386, i486, i586 or i686 represents 32 bit kernel machine.

Step 4: Start Libvirt by executing command as shown in Figure 7.4 and Figure 7.5.

```
hpc-client@hpcclient-H61MLV3:~$ sudo service libvirtd start
hpc-client@hpcclient-H61MLV3:~$ sudo update-rc.d libvirtd enable
hpc-client@hpcclient-H61MLV3:~$ service libvirtd status
```

FIGURE 7.4: Starting LibVirt Service.

```
○ libvirtd.service - Virtualization daemon
   Loaded: loaded (/lib/systemd/system/libvirtd.service; enabled; vendor preset:
   Active: active (running) since Mon 2019-06-03 14:41:35 IST; 6min ago
     Docs: man:libvirtd(8)
           https://libvirt.org
 Main PID: 6210 (libvirtd)
    Tasks: 19 (limit: 32768)
   CGroup: /system.slice/libvirtd.service
           ├─6210 /usr/sbin/libvirtd
           ├─6772 /usr/sbin/dnsmasq --conf-file=/var/lib/libvirt/dnsmasq/default
           └─6773 /usr/sbin/dnsmasq --conf-file=/var/lib/libvirt/dnsmasq/default
```

FIGURE 7.5: Checking Status of LibVirt Service.

7.4.3 Creating a virtual machine

Step 5: Run the virt-manager as shown in Figure 7.6.

```
hpc-client@hpcclient-H61MLV3:~$ sudo virt-manager
```

FIGURE 7.6: Executing the Virt-Manager.

Step 6: On execution of step 5, a window will open as shown in Figure 7.7.

FIGURE 7.7: Creating VM.

Step 7: Goto fle and create VM as shown in Figure 7.8. An ISO image of required OS, need to be there in your machine.

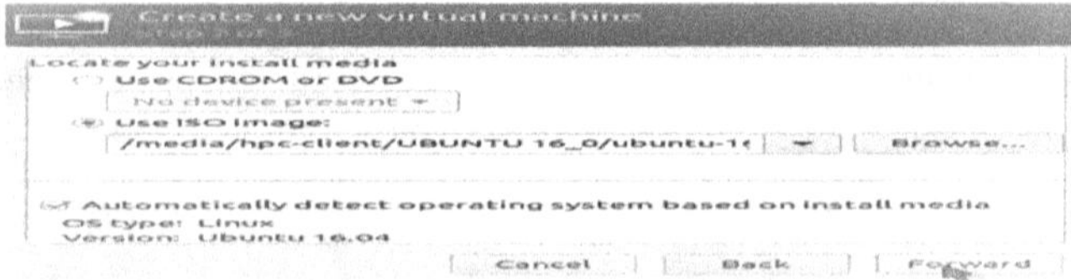

FIGURE 7.8: Selecting ISO VM.

Step 8: Configure the VM as shown in Figure 7.9. Select the required configurations of kernel, RAM and storage.

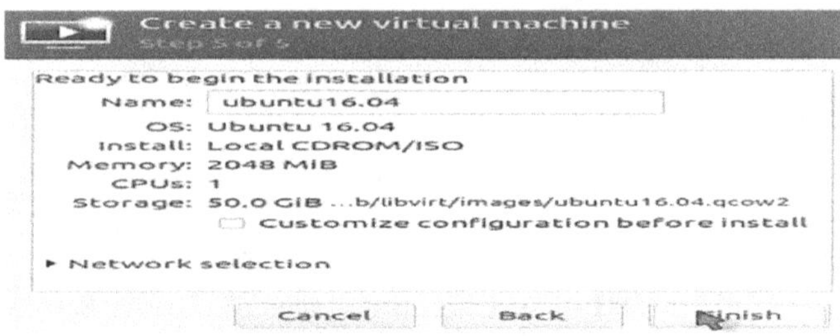

FIGURE 7.9: Configuring VM.

Step 9: Press finish and create VM as shown in Figure 7.10. It might take sometime.

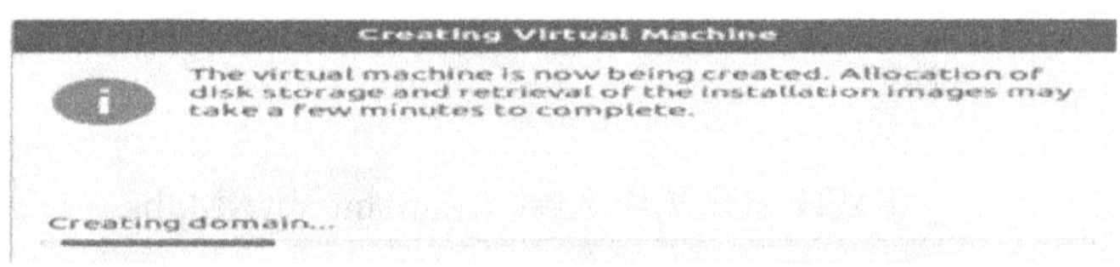

FIGURE 7.10: Installing VM.

7.4.4 Install LibVMI tool and necessary dependencies

Step 10: Install LibVMI [164]. Install the following dependencies:

- libvirt (KVM only), libxc (Xen only), libxenstore (Xen only)
- yacc or bison
- lex or flex
- glib version 2.16 or newer
- libtool, m4, automake

Step 11: LibVMI uses the standard GNU build system. To compile the library, run he following command, as shown in Figures 7.11, 7.12, and 7.13.

```
hpc-client@hpcclient-H61MLV3:~/Desktop/libvmi-0.10.1$ ./autogen.sh
libtoolize: putting auxiliary files in '.'.
libtoolize: copying file './ltmain.sh'
libtoolize: putting macros in AC_CONFIG_MACRO_DIRS, 'm4'.
libtoolize: copying file 'm4/libtool.m4'
libtoolize: copying file 'm4/ltoptions.m4'
libtoolize: copying file 'm4/ltsugar.m4'
libtoolize: copying file 'm4/ltversion.m4'
libtoolize: copying file 'm4/lt~obsolete.m4'
configure.ac:93: installing './compile'
configure.ac:6: installing './missing'
examples/Makefile.am: installing './depcomp'
hpc-client@hpcclient-H61MLV3:~/Desktop/libvmi-0.10.1$ 
```

FIGURE 7.11: Installing LibVMI.

```
hpc-client@hpcclient-H61MLV3:~/Desktop/libvmi-0.10.1$ ./configure
checking for a BSD-compatible install... /usr/bin/install -c
checking whether build environment is sane... yes
checking for a thread-safe mkdir -p... /bin/mkdir -p
checking for gawk... no
checking for mawk... mawk
checking whether make sets $(MAKE)... yes
checking whether make supports nested variables... yes
```

FIGURE 7.12: Installing LibVMI.

```
Feature       | Option                  | Reason
--------------|-------------------------|---------------------------
Xen Support   | --enable-xen=no         | missing xenstore
Xen Events    | --enable-xen-events=no  | no
KVM Support   | --enable-kvm=yes        | yes
File Support  | --enable-file=yes       | yes
--------------|-------------------------|---------------------------
```

FIGURE 7.13: Installing LibVMI.

Step 12: Configuring LibVMI and Finding Offsets of monitored VM, as shown in Figures 7.13 to 7.15.

```
hpc-client@hpcclient-H61MLV3:~/Desktop/libvmi-0.10.1$ cd tools
hpc-client@hpcclient-H61MLV3:~/Desktop/libvmi-0.10.1/tools$ ls
linux-offset-finder  performance  pyvmi  qemu-kvm-patch  vmifs  windows-offset-finder
hpc-client@hpcclient-H61MLV3:~/Desktop/libvmi-0.10.1/tools$ cd linux-offset-finder/
hpc-client@hpcclient-H61MLV3:~/Desktop/libvmi-0.10.1/tools/linux-offset-finder$ sudo su
[sudo] password for hpc-client:
root@hpcclient-H61MLV3:/home/hpc-client/Desktop/libvmi-0.10.1/tools/linux-offset-fincer# ls
findoffsets.c   findoffsets.mod.c  findoffsets.o  modules.order   README
findoffsets.ko  findoffsets.mod.o  Makefile       Module.symvers
```

FIGURE 7.14: Finding Offset (part 1).

```
root@hpcclient-H61MLV3:/home/hpc-client/Desktop/libvmi-0.10.1/tools/linux-offset-finder# make
make -C /lib/modules/4.15.0-50-generic/build M=/home/hpc-client/Desktop/libvmi-0.10.1/tools/lin
ux-offset-finder modules -lcsys
make[1]: Entering directory '/usr/src/linux-headers-4.15.0-50-generic'
Makefile:976: "Cannot use CONFIG_STACK_VALIDATION=y, please install libelf-dev, libelf-devel or
 elfutils-libelf-devel"
  Building modules, stage 2.
  MODPOST 1 modules
```

FIGURE 7.15: Finding Offset (part 2).

```
root@hpcclient-H61MLV3:/home/hpc-client/Desktop/libvmi-0.10.1/tools/linux-offset-finder# insmod
 findoffsets.ko
root@hpcclient-H61MLV3:/home/hpc-client/Desktop/libvmi-0.10.1/tools/linux-offset-finder# 
```

FIGURE 7.16: Finding VM Offset (part 3).

Step 13: Now go to the 'Syslog' file and scroll down to the current timestamp, you will find the offsets, as shown in Figure 7.17.

```
Jun  4 11:02:56 hpcclient-H61MLV3 kernel: [ 5223.816171] findoffsets: module verification fa
signature and/or required key missing - tainting kernel
Jun  4 11:02:56 hpcclient-H61MLV3 kernel: [ 5223.816293] Module FindOffsets  loaded.
Jun  4 11:02:56 hpcclient-H61MLV3 kernel: [ 5223.816293]
Jun  4 11:02:56 hpcclient-H61MLV3 kernel: [ 5223.816296] [domain name] {
Jun  4 11:02:56 hpcclient-H61MLV3 kernel: [ 5223.816297]     ostype = "Linux";
Jun  4 11:02:56 hpcclient-H61MLV3 kernel: [ 5223.816298]     sysmap = "[insert path here]";
Jun  4 11:02:56 hpcclient-H61MLV3 kernel: [ 5223.816299]     linux_name = 0xa50;
Jun  4 11:02:56 hpcclient-H61MLV3 kernel: [ 5223.816300]     linux_tasks = 0x7a8;
Jun  4 11:02:56 hpcclient-H61MLV3 kernel: [ 5223.816301]     linux_mm = 0x7f8;
```

FIGURE 7.17: Finding VM Offset (part 4).

Step 14: Now, copy the offset information from syslog (highlighted in blue) and paste in /etc/libvmi.conf as shown in Figure 7.18

```
ubuntu16.04 {
     ostype = "Linux";
     sysmap = "/home/hpc-client/Downloads/System.map-4.15.0-29-generic";
     linux_name = 0xa50;
     linux_tasks = 0x7a8;
     linux_mm = 0x7f8;
     linux_pid = 0x8a8;
     linux_pgd = 0x50;
}
```

FIGURE 7.18: Configuring LibVMI configuration file.

Step 15: Now run the following command 'sudo ./examples/process-list.c ubuntu16.04' and extract the process list of VMs from KVM as shown in Figure 7.19

```
1532    [0x39cfd57c3]:    ksoftirqd/0
1533    [0x39afd5018]:    amazon
1534    [0x33cfd5218]:    LibreOfficeCalc
1536    [0x33cfd50c9]:    Help
1537    [0x33cfd50c8]:    watchdogd
1538    [0x37cfd50bc]:    cpuhp/0
1540    [0x37cfd5027]:    firefox
1541    [0x39cfd50c1]:    watchdog/1
1546    [0x34cfd50c8]:    LibreOfficeWriter
1548    [0x28cfd73a2]:    UbuntuSoftware
```

FIGURE 7.19: Extract the VM process list from KVM using LibVMI.

7.5 Conclusion

Security breach is a nightmare for every system administrator. Attackers procure access to the system and thieve sensitive information, locking critical files, or leaking private information outside. To recover from such attacks becomes very strenuous for the target. Virtualization acts as the major innovation for empowering the sharing of resources in the cloud environment. The security of classified data stored in cloud is a crucial apprehension for assurance of the user. Thus providing security from attackers and hackers has become a huge challenge. This chapter provides a summary of some security tools for various types of attacks whether on the network or host. Along with security tools, this paper also provides description of some attack tools. Each attack tool has its own strategy of doing attacks. There is a need to provide efficient prevention techniques against attacks. There are a number of tools for security against attacks. However, no one tool exists that can launch a variety of attacks and provide safety against all possible attacks causing disruption. This paper provides with a list of distinguished attack tools that can be used for generating huge data sets and hence can be used for research purposes.

7.6 Questions

Fill in the blanks

1. Mark the correct option. The network-layer attack tools are

   i HULK
   ii Low Orbit ION cannon
   iii XOIC
   iv All of above

2. The Hypervisor-layer can be attacked by

   i Blue Pill
   ii Subvirt Pill
   iii BlueHill Pill
   iv None of these

3. Mark the incorrect statement

   i Abel is a password recovery tool for the Microsoft Operating Systems.
   ii Open Source HIDS SECurity (OSSEC) is an open-source, free host based intrusion detection system (HIDS).
   iii Cloud Custodian is an engine of rules that manages cloud accounts which are public.
   iv Nitro is the introspection tool for Xen only.

4. Mark the correct statement in the followings

   i Dsniff is a collection of tools which are used for password sniffing and network traffic analysis.
   ii A worm is a program that is independent and creates its multi- ple copies itself across the network.
   iii Drakvuf is a virtualization-based agent less black-box binary analysis system.
   iv All are correct

5. Mark the correct statement. Rekall is a/an

   i Advanced forensic and response framework
   ii Network traffic capturing tool
   iii Static memory acquisition tool
   iv None of these

Short-Answer Questions

1. What is the difference between VM-level and VMM-level security tools?
2. Explain DRAKVUF and LibVMI security tools with their key features.
3. Compare various attack tools in terms of different parameters.

Long-Answer Questions

1. Draw the taxonomy of security tools. Discuss any two security tools in each category with features.
2. Draw the taxonomy of attack tools. Discuss any two attack tools in each category with features.

Chapter 8

Virtual Machine Introspection and Hypervisor Introspection

8.1 Introduction

Virtual machine plays an important role in building high assurance system as it capable of isolating the multiple environments. Each of the environments may have different security concerns. IDS play an important role in identifying and detecting the security threats of the virtual environments. Some researchers are working toward IDS approaches. However, hackers are also working toward creation of advanced malware with same space. The advanced malicious codes are so well-advanced that they can easily detect the security detection components.

The existing cloud security frameworks which deploy the IDS inside VMs may prone to subversion attacks. The advanced malware try to hide itself from the security analyzer and can even sense the analysis environment. The hidden malware and rootkits may get bypassed by security tools running inside VM. In addition, the current OSes does not provide a strong isolation which makes the guest based security approaches weak and subjected to attacks. Hence, researchers proposed solutions to move the security analysis outside the guest VM and isolate them from the monitored environment.

However, the TVM-layer security mechanisms cannot be directly implemented at the VMM-layer. Hypervisor can only view the bits and bytes of guest OS along with hardware registers. There exists no high level semantics of processes, data structures or files. Hence, the TVM-layer security cannot work in the raw view of guest OS. This problem is called as 'semantic gap problem'. The concept of capturing the abstract information, reconstructing it and getting the high-level view of guest OS information (memory, hardware states, etc.) is called "Bridging the semantic gap".

The advanced security mechanisms can be used to bridge that gap and provide similar detection capabilities of traditional approaches outside the monitored VM with much better attack resistance. Since last few years, researchers have well investigated the advanced security solutions, called introspection. Introspection also provides the capability to monitor the VM at the hardware-level as well. Moreover, introspection enables to capture and analyze the network traffic outside the VM by performing the network introspection at virtual networking layer. It makes sure that the traffic going out-of-VM or traffic coming toward VM should not be vulnerable and prone to attacks.

The introspection approaches which try to detect VM-level attacks from hypervisor are called as virtual machine introspection (VMI) approaches. They assume that

DOI: 10.1201/9781003004486-8

hypervisor is a trusted domain. Some of the popular VMI approaches are Livewire [241], Maitland [202], VMGuard[168], and VAED[169], etc. These approaches are deployed outside the VM on or above the hypervisor or hardware-layer which is the more trustworthy place when compared to VMs. However, along with VM security, hypervisor security is also important. Hypervisor introspection mechanisms provide advanced security approaches, capable to detect hypervisor attacks. The introspecting component is either placed at the hypervisor or below the hypervisor. HyperSentry [285] and CloudVisor[215], etc. are some of the popular HI approaches.

In this chapter, we mainly focus to provide the followings: (i) A detailed classification and technical description of various VMI techniques with their comparative analysis; (ii) A detailed classification and technical description of various HI techniques with their comparative analysis; (iii) A case study of VMI security tool.

Let us first discuss VMI approaches in details:

8.2 Virtual Machine Introspection (VMI)

VMI approaches are designed to analyze the VM code from an isolated region outside monitored VM. It makes use of hypervisor features and introspection libraries such as LibVMI [164]. The introspecting libraries use guest OS symbol file such as system.map file to extract the information related to memory regions. These libraries provide the high-level view of VM states information. In most of the VMI approaches, the security analyzer runs in the security VM (Dom0) and all its components are completely isolated from the monitoring VM. However, in some of the VMI approaches, a small security code, also called as trampoline code is placed inside the monitoring VM. This code keeps on monitoring the VM events and communicates with security VM modules. The security analyzer analyzes the VM information and if any suspicious call is detected, an alarm is generated to the administrator. The conceptual diagram for VMI approaches is shown in Figure 8.1. A cloud administrator may use VMI for monitoring the VMs from the security VM and analyzing the user behavior. VMI-based IDS tools are deployed at the hypervisor-layer or inside the privilege domain. These tools can detect attacks which can be launched against VMs from outside.

All the VMI-IDS, security mechanisms have been classified into following categories based on: (i) VM-State Information, (ii) VM Hooks, (iii) Hypercall verification, (iv) Guest kernel debugging, (v) VM interrupt analysis. A brief comparative discussion of various approaches is highlighted in Table 8.1. Let us now discuss each of the categories in detail.

8.2.1 VM hook based

VM hook-based approaches employ an In-VM agent (called hooks), which is injected inside the guest OS of monitored VM. The hooks are mainly kernel modules which intercept the events and send the VM information to the VMI-IDS running in security VM. There is a need to modify the guest OS to inject these hooks which makes these approaches less attack resistant.

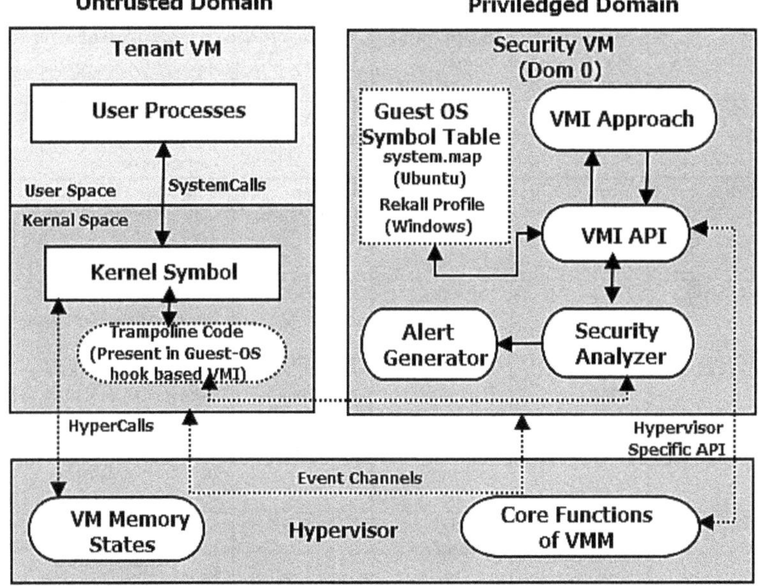

FIGURE 8.1: Conceptual working of VMI approaches.

Kourai and chiba [288] proposed a VM introspection based framework to detect malware attacks. The first VM (monitored VM) runs user applications. The monitoring VM runs the security monitor placed in the same host. The first VM is also called as introspected VM. The frames are sent and received using a mirroring software. One of the introspection frameworks [163] places security analyzer in the untrusted VM. In their security approach, memory protection and hardware virtualization feature is utilized to create VMM protected address space. The log events are maintained in the secure place and accessed in controlled way. A sample prototype is implemented in KVM servers having Intel-VT support.

Lares [287] is another introspection architecture which also makes use of two VMs. The machine which is under monitoring is called as guest VM which executes the user applications. Some security hooks are injected inside this VM. They gather and pass the information about the VM activities (events) to the hypervisor. The hypercalls, trapped by hypervisor are passed to the analysis VM (second VM). This VM runs the core security functionalities to detect and analyze the events. The security tool then enforces the decision to the VM under monitoring.

8.2.2 VM-state information based

VM-state information is very helpful in extracting the useful VM data for analysis. A state of a VM is defined by CPU registers, I/O states and memory information, etc. The VMI libraries/tools help in gaining that information and converting it into semantically meaningful form.

TABLE 8.1: Summary of VMI techniques

VMI Framework	Supported Hypervisor	Approach used	Supported Guest OS	Limitations
XenAccess [165]	Xen	VM State Access	Linux and-Windows	- Kernel rootkits can attack - obsoleted
Livewire [184]	VMware Workstation for Linux 3.1	VM State Access	Debian GNU/Linux	-can expose to rootkits - fails for unknown attacks
Ether [243]	Xen 3.1.0	Interrupt Forcing	Windows XP SP2	- limited to instruction-level traces -high overhead
Nitro [224]	KVM 0.12.4	Interrupt Forcing	Linux and windows	- prone to rootkits - supports KVM system only
DRAKVUF [166]	Xen 4.6	Kernel Debugging	Windows 7	- limited to kernel function trapping -overhead is high
SPEMS [286]	Xen 4.4.1	Kernel Debugging	Windows	- limited user-level functions functions
Xenini [175]	Xen 4.0.1	Interrupt Handling	Linux	- storage requirement is high - fails for rootkit attacks.
Collabra [173]	NA	Hypercall authentication	NA	- lacks implementation -prone to traffic congestion
SIM [163]	KVM	Guest OS hook	Windows XP SP2	- fails for rootkits - memory protection is weak
Lares [287]	Xen 3.0.4	Guest OS hook	Windows XP SP2	- Rootkits can evade - supports only older version of VMM

Payne et al. [165] proposed XenAccess VMI architecture which provides support for Xen hypervisor. The library provides the functionality for virtual disk monitoring from VMM. It creates semantic-aware abstraction to provide an easy access to user VM from privileged domain (Dom0). LibVMI [164] is the extended framework of XenAccess which supports more features than XenAccess . It provides introspection functionality for both Xen and KVM hypervisor. It is very useful library which is used by various VMI-based security tools.

Maitland [202], is one of the light-weighted approaches which addresses security in the paravirtualization. The monitoring VM and privilege VM are the main VMs which play significant role in complete framework. The former runs user applications and later analyses the information about the monitoring VM. Let us briefly understand its key functions. The approach keeps on monitoring the dirty bit's flag status corresponding to memory pages of the active processes. Any memory update is notified to the analyzer. If the process is trustable then only it is allowed to execute. If a page fault is notified, then again the framework ensures if it is generated by some legitimate process. A memory snapshot of suspicious processes which generated such page faults is taken and notified to the admin so that some action can be taken in prior.

Tang et al. [289] proposed RansomSpector that is a VMI-based technique for dealing with crypto-ransomware that runs in the operating system. RansomSpector resides in the VMM layer and monitors network activities and the filesystem to detect ransomware. This technique is useful in virtualized and cloud environments as service providers secure their customers. It obtains better precision and earlier ominous. They implemented RansomSpector's prototype and gathered 2117 samples of the ransomware for evaluation and they detected 771 ransomware samples that are based on network activities pattern and file I/O access. For the experiment, they used 64 bits Windows 7 as a virtual machine operating system. The result shows better performance to detect ransomware with 0 false positives but with less than 5% performance overhead.

A network introspection based malicious network packet detection approach [290] is a VM-sate information based mechanism to detect the network attacks at the VMM-layer in cloud. It captures and analysis the traffic at the Dom0 of the Xen based Compute Servers and the centralized Network Server. The execution flow is shown in Figure 8.2. Initially, a traffic validation module is executed to validate whether the traffic is coming from the source VM in running on the top of the hypervisor layer. The module is executed at the privileged domain of Xen-VMM which makes use of *xenstore* and *XenAPIs*. If a traffic is found to be IP or MAC spoofed; an alert is raised to the Cloud administrator as a primary defense machanism. If network traffic is not spoofed; it is then passed to the secondary module deployed outside the VM for behavior analysis using machine learning classifier, particular Random Forest (RF) is used. The behavior analysis module is trained over known attack patterns which makes it a better choice to detect known attacks and their variants. The proposed approach has been validated with latest datasets (UNSW-NB and ITOC) and results seem to be promising.

8.2.3 Hypercall verification based

Just like system calls, hypercalls are also a software trap or interrupts from VM layer to the hypervisor kernel. In order to execute the privileged operations, hypercalls are invoked by para-virtual machines (PVM). Let us discuss some of the hypercall based introspection based approaches below:

Collabra [173] is an introspection based conceptual framework for VM security in virtualization environment. It filters the communication between privileged VM and monitoring guest VM. It is based on checking the integrity of hypercalls. A collaborative defense mechanism is provided to prevent the security threats at the hypervisor layer. The key component of collabra is under control of admin and runs alongside of privileged VM. The security components communicate with other collabra components in different machine via a logical control channel. On execution of a critical event, collabra instance starts the monitoring and checks the integrity of hypercalls. It has got mainly two key security modules: hypercall integrity checker and origin checker. A message authentication code (MAC) is maintained along with specified policies for each of the hypercall. It is later used to verify each of the hypercall issues during program execution. In case, if a call is completely unknown, the other security instances of collabora, deployed in different servers are consulted for correctly classifying such calls. A legitimate call passes all security checks and executed in normal way. However, the calls invoked by unknown programs are marked as untrusted. The information is shared with other collabra instances. The paper does not provide any prototype implementation of the proposed work.

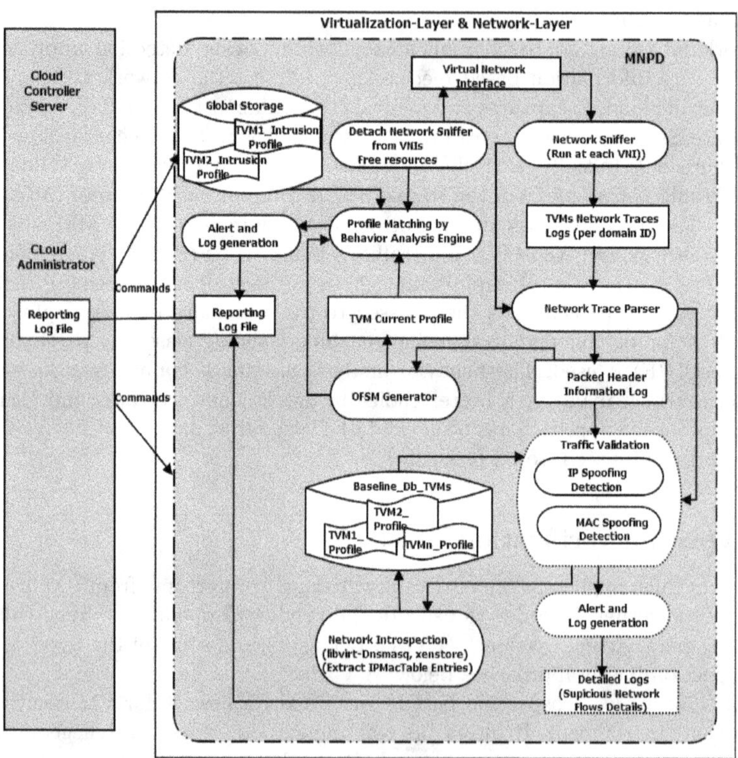

FIGURE 8.2: Out-VM malicious network packet detection based on network introspection).

8.2.4 Guest OS kernel debugging based

Some of the VM instrospection approaches are based on kernel debugging mechanisms in which the kernel data is analyzed by extracting the important kernel logs. The break point injection is one such technique in which the break points are inserted at specific kernel symbols so as to trace the execution of system calls and necessary information is extracted about the kernel and user program interaction. Let's discuss some of such techniques:

DRAKVUF [166] is one of the fabulous introspection tools which is based on the breakpoint injection technique and helps in detecting rootkits and some other malware attacks. The file system is accessed by kernel's heap memory. LibVMI is employed to extract the basic information of VM proceeses, running inside the monitored VM. It traps the kernel functions by employing the break point injection method. The rootkits and malicious drivers can also be detected by this security tool. In order to inject the breakpoint at the specific locations, it is essential to have the information related to the kernel functions locations. Hence, the kernel debugging is required to have access to such information by making use of forensic tools such as Rekall [283]. The logs generated by DRAKVUF, represents the full system call tracing on execution of the monitored program. It can also detect the kernel object manipulation attacks. This framework is limited to the Xen hypervisor and also it does not support the hypercall tracing features.

SPEM is introduced by some researchers [286] after DRAKVUF to detect the anomalies using hypervisor support. It extends some of the features of it. DRAKVUF performs the kernel level function trapping; however SPEM extends the kernel level tracing and adds the support of user-level function tracing. The malware programs that bypass the user-level code, has got different features than the malware which attacks the kernel-level function calls. Hence, SPEM does some modification in the existing tool to trace both type of malware. They have classified user level functions and kernel level functions. The tracing of trapping of program execution happens in the same way. However, all types of functions are considered for monitoring. The main and call back function of DRAKVUF is changed. The syscall traces are obtained. Similar to existing tool, it also does not support the hypercall tracing mechanisms.

Zhan et al. [291] proposed a VMI approach for monitoring the kernel objects that reduces overhead due to monitoring at the page-level. The key concept is to automatically place the kernel objects at a more secured region and then securing that region from attacks. The new memory region holds the kernel objects to be secured; the need of monitoring for entire kernel objects is removed. There is no overhead at runtime to the associated kernel service. They used Xen hypervisor and Ubuntu 16.04 x64 and as target virtual machine for the experiment. The result shows this system has little overhead and monitor target kernel objects efficiently.

Mishra et al. [169] proposed a memory introspection based approach in which the security monitor is deployed on the top of VMM kernel inside the privileged VM as shown in Figure 8.3. It performs various functions such as system call logs extraction using break-point injection based introspection mechanism, parsing of system calls and generating a System Call Dependency Graph (SCDG) for each of the execution trace. Each edge of SCDG is marked with a probability value calculated using Markov property. The feature vectors are generated by traversing each possible

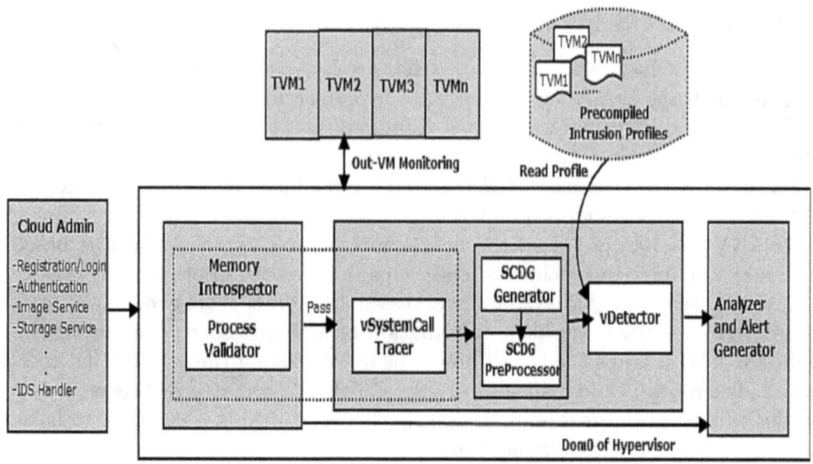

FIGURE 8.3: VAED: basic security architecture.

path from one system call to another system call. The approach provides promising results using some evasive dataset.

VMShield [292] is an advanced introspection-based cloud security architecture, which is based on kernel debugging-based VMI approach and uses DRAKVUF [166] library to extract the program execution traces of applications running in the VM at the Xen hypervisor. The introspection tools are deployed inside the privileged domain (Dom0) of Xen. The use of memory introspection makes the approach robust over traditional security approaches. VMShield is based on the assumption that Dom0 domain is a secure domain and cannot be breached. The collected traces are parsed and converted into feature matrix. The important features are selected using the binary particle swarm optimization (BPSO) approach which extracts the important features and makes the dataset optimal. It further invokes the decision mode which is trained over random forest classifier. The decision model detects the abnormal behavior in the program execution traces as shown in Figure 8.4. The approach provides good results with less overhead over the UNM [231] and Barecloud dataset [293].

8.2.5 VM interrupt analysis based

Interrupt based introspection approaches trap the execution of kernel functions, generated in the target machine. Let's describe each of such approaches below:

These category of techniques are based on forcing the interrupts by injecting the hardware hooks (e.g., setting and unsetting the specific register values and exceptions, etc.). They derive the VM specific information such as specific register value, processor context information for doing the active analysis on VM activities.

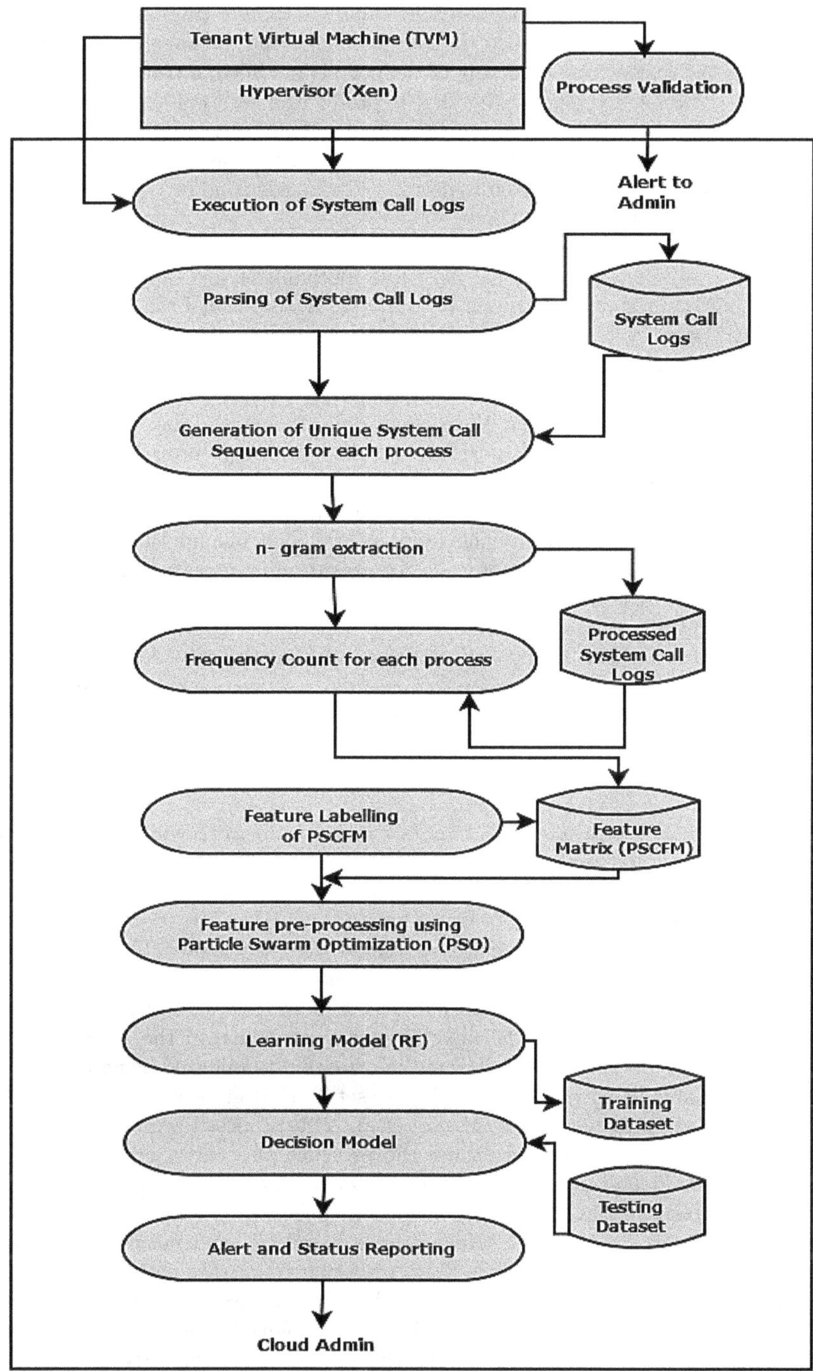

FIGURE 8.4: VMShield: basic security architecture.

Ether [243] is one the introspection based analysis tool for trapping the execution of the VM programs. It is based on the idea of monitoring the page modifications in by checking the register values. If any of the page is modified, a trap is generated to extract the dynamically generated codes. It produces a series of system calls as an output. It makes use of fast system call method software interrupts to produce the execution tracing logs. The major limitation with this opens source tool is that it only supports older version of Xen hypervisor. There are is no further update being done in the design since a long time.

Nitro [224] is another system call tracing framework based on interrupt forcing mechanism in virtual environment. All three mechanisms are supported by Nitro for collecting the system call traces. It supports: system calls based on interrupts, SYSENTER based system calls `SYSCALL/SYSRET` based system calls. In case of syscall interrupts, interrupts are generated on execution of `SYSRET` or `SYSCALL`. These are trapped by hardware support (Intel-VTx support). For the user interrupts in the machine, nitro starts extracting the system calls to extract the behavior of the programs under monitoring. A performance degradation is produced in between 5–54%. It is limited to KVM hypervisor which is its major limitation.

ShadowContext [208] is one of the intrusion prevention frameworks to detect attacks in virtual environment. The fundamental idea behind this framework is redirecting the some selected system calls for execution in a trusted machine (Dom0 VM). This is how sensitive system calls can be prevented from any malware attack. However, the framework injects some security code in the guest VM as well which is under monitoring. The security of such code is also very important for the successful implementation of redirection and execution of system call at the specified location. The security is achieved using the EPT features. The unsuccessful implementation can even the crash the monitored machine and hence the correctness in implementing this approach should be as high as possible. It is validated in KVM based systems.

Xenini IDS [175] is a interrupt handling based framework which generates the system call traces and performs the analysis on them. The framework is designed for Xen VMM architecture. The trapping happens on arrival of systemcall interrupts (0X80) and the control gets transferred to XenIDS. The tool is deployed in the Dom0 VM. The id of system call is extracted from the value of %eax register. A security patch (Xenini) is added as a patch to the hypervisor for gaining access to required information about the VM. This patch interacts with XenIDS and passes the gather information to it. The libxc library is used to read the intercepted information from hypervisor. The extracted information is analyzed using the STIDE [230] and validated using very older UNM dataset [231]. It gives an overhead of 5–6% during prototype implementation. Mishra et al. [214] proposed a process validation approach to detect rootkits and ensure the presence of security critical processes in KVM based virtualization server. They extended the security architecture and named, KVMInspector [294] to detect intrusions in KVM-based cloud environment as shown in Figure 8.5. The KVMInspector is based on the concept of interrupt-handling based VMI and uses Nitro [224] VMI tool to extract program, execution logs of VM programs at KVM VMM. The In-VM logs are extracted using strace utility in Linux VM. The LibVMI [164] library is deployed at KVM to perform the program validation to ensure that all security processes are correctly executing inside the VM. The monitored logs are parsed and converted into feature matrix using bag of n-gram approach and important features are extracted using RFE. It optimal

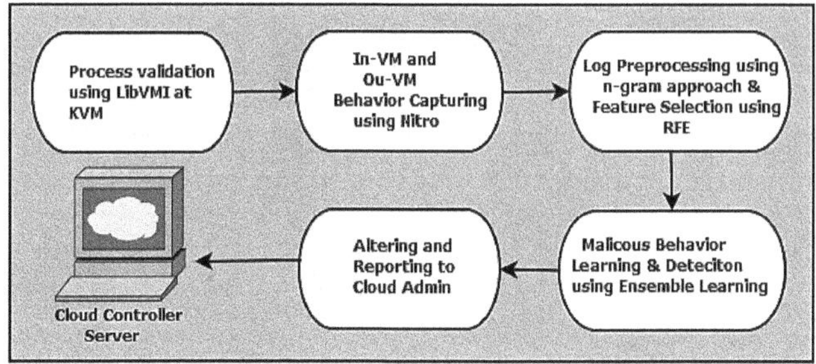

FIGURE 8.5: KVMInspector: basic security architecture.

features set is passed by heterogenous ensemble classifier in which results of each classifier passed through voting classifier at the end to make common decision. The approach is validated using UNM [231] and University of New California dataset [293] and achieves promising results.

8.3 Hypervisor Introspection (HVI)

All of the above mentioned VM introspection approaches assume that hypervisor is a trusted domain. However, if hypervisor security is breached, then all VMs running in upper layer will also be compromised. Hence, securing the hypervisor is also as important as securing the VMs.

Introspection of the VMM/hypervisor and gaining access to the useful hypervisor specific information such as hypercall logs, memory regions, control and non-control flow data, is called as hypervisor introspection (HVI). These techniques are designed to be deployed below the layer of VMs. Lets us classify them in five categories: (i) Nested Virtualization, (ii) Code Integrity Checking using hardware-support, (iii) Memory Integrity Checking using Hardware/Software Support, (iv) Revisiting the VMM Design, (v) VM-Assisted Hypervisor Introspection, as shown in Figure 8.6. Each of the category is explained in detailed below:

8.3.1 Nested virtualization

Let's discuss some of the nest virtualization-based hypervisor security approaches.

GuardHype [295] is one of the nested virtualization-based security frameworks in which a security layer is placed below the hosted hypervisor. Some of the dangerous rootkit attacks (VMBR) can target hypervisor and thereby considered as a potential threat to virtualization. The entire control of guest OS kernel is taken over by VMBR rootkit, it pretends itself as a layer below VM, i.e. hypervisor-layer. Such types of attacks are known as hyperjacking attack. Now, the security framework analyzes the VM behavior to check if there is some hyperjacking activity and does the monitoring of events between VM and hardware. It is added as a thin layer

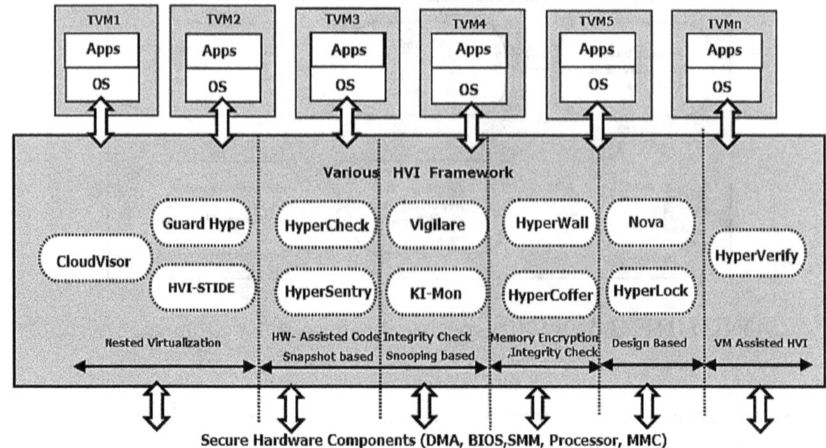

FIGURE 8.6: Security frameworks for hypervisor introspection.

between hardware and monitored hypervisor. It compares the Virtual Machine Control Structure (VMCS) value with the values set by host OS. If a attacker is successful in gaining control on guest OS, the host OS's VMCS value will change. The major challenge with such approaches is that the amount of overhead introduced in the system that affect the overall performance of the system.

Cloudvisor [215] is another framework for securing the hypervisor in which the concept of nested-virtualization is used. The key idea behind this framework is that the security functions of main hypervisor can be separated from the main layer in a more protected region. The framework introduces a thin layer below the main hypervisor layer. It is designed to protect the hypervisor from attacks, targeted from VMs. It ensures that privacy and integrity of the hypervisor layer is maintained. The interaction between VMM and guest OS is monitored by the thin layer. It maintains the CPU state information about the VM memory pages. CloudVisor handles the VM_Exit events which are validated and forwarded to hypervisor if founds valid. The contexts of CPU registers are saved and address translation is manipulated to isolate the VMM memory from security tool. The extended page table scheme is used to ensure that VMs are isolated. The mapping/remapping is done via hypervisor. The system overhead is 16.8%. It cannot detect the physical attacks such as cold boot attacks [144].

Hypercalls vulnerabilities can also be exploited just as attackers exploit the systemcall vulnerabilities. HInjector tool [276] is one such attacking tool that injects the malicious hypercall attacks. The insertion or tampering in the hypercall (modifying the parameters or calling unwanted sequence, etc.) can be done by attackers with an intention to cause harm. Some authors [249] proposed the hypercall based security framework to detect such attacks as shown in Figure 8.7. These calls cannot be traced at the guest level. There are hypercall tracer plugin which can be used at the VMM layer to trace and intercept the hypercalls. In their framework, they deployed the nested virtualization layers; a secure VMM layer (L0) is injected on the top of system-level hypervisor (L1). The L0 is a new version when compared to L1 and is designed to protect the base layer L1. The main components are: interceptor

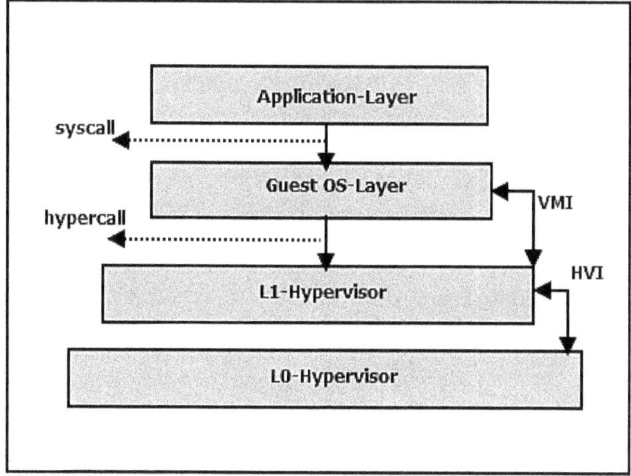

FIGURE 8.7: Framework for nested-virtualization.

which intercept the hypercalls and analyzer which further analyses them. These hypercalls are analyzed by the STIDE [230] approach for checking the presence of any suspicious hypercalls. The system system also faces some overhead. Pattern matching is used to perform the analysis which can be evaded by modern malware. These approaches cannot detect the side channel attacks.

8.3.2 Code integrity checking using hardware-support

A new set of vulnerabilities are added when the hypervisor is introspected from an additional layer, added above the hardware. Hence, it generates the need of having more advanced solutions such hardware-supported solutions. Below, we describe two such categories: Snooping-based and snapshot-based.

Snooping-based

These types of approaches are transient to attacks which can cause harmful operations during snapshots. No trace is left out in memory. In order to address this, frequent snapshots can be taken or some randomized intervals can be considered for such operations. However, system overhead could be increased in such scenarios. Hence, some researchers proposed even-triggered based techniques.

Vigilare [245] is a kernel code integrity checking based framework which is capable in detecting transient attacks and some hardware attacks as well. The snooper and verifier are the two key design components of this framework. The memory bus traffic of the host under monitoring is collected by snooper. The filtered traffic is now further verified by the later design component to check the integrity of memory contents. A selective bus traffic algorithm is used further to perform filtering of traffic. The system has been validated in a machine with Leon3 processor, linux kernel ver. 2.6. The immutable sections of kernel are monitored by this framework. It achieves in lesser performance overhead than snapshot-based approaches. The mutable kernel sections are not monitored by this approach.

KiMon [296] is another security framework which helps in monitoring the kernel objects (mutable) and tries to find the anomalies specially rootkits. Mutable objects can be modified at the kernel level. Initially, the traffic of memory bus (write traffic) is snooped and compared against the whitelist registers. DMA is used to perform the read operations. Incase if a mistmatch is detected, the event signals are generated and forwarded to kernel level integrity verifier. The semantics is also verified with the kernel objects. A modification in the kernel structure is required in this framework. The framework is validated and achieves 100% detection rate for detecting kernel level attacks.

Snapshot-based

In this category of approaches, memory and CPU snapshots are taken at some time intervals. Various hardware features such as system management mode (SMM) can be used to this purpose.

Copilot [297] is one of the kernel code integrity checking-based frameworks which helps in detecting various kernel level rootkit attacks. The copilot monitor is deployed in the host machine (inside the PCI card). A remote communication with the copilot is done by admin station. The kernel integrity is preserved by using the MD5 hashes. The critical memory regions are monitored for the presence of any malicious processes. The virtual addresses of linux kernel symbols are extracted using `system.map` file. The security tool is polled by the admin station via PCI bus. The detection rate is 12 rootkits per 30 sec. This is prone to timing threats.

Hypersentry [285] is one of the novel VMM integrity checking-based frameworks. An out-of-band channel is configured by using an intelligent interface. On target machine, system management interrupt (SMI) is triggered which invokes system management mode (SMM). The interface is implemented on firmware. The context information (like CR3 value, VMX root operation, etc.) of the hypervisor is acquired by using a fallback technique. Once the context information is retrieved, SMI handler is invoked to secure the information. It provides the 2.4% system overhead within 8 seconds.

Hypercheck [116] is a hypervisor integrity monitor which helps in checking the integrity of softwares, booted after BIOS such as VMM, OS, etc. The current state of CPU and memory are extracted in form of snapshots in the SMM mode of CPU. SMI is used to invoke the SMM and the current state of the system is then saved in SMRAM, a separate address space in RAM. The register monitoring and memory acquisition components are placed inside the VM under observation. The analyzer is deployed inside the security VM which performs analysis. Once data is acquired, the hashing technique is applied and then the encrypted data is transferred to security machine. Next component compares the memory snapshots (stored prior) and ensures the integrity. The CPU register values are fetched from SMRAM and are validated. The monitoring machine does the further analysis on the reports sent by the first two modules. The system performance is 2% for 5 seconds of sampling period.

8.3.3 Memory integrity checking using hardware/software support

The above mentioned techniques are hardware dependent and requires only hardware support of SMM, DMA and PCI etc for checking the integrity of the memory. Hence, the software assistance-ship is also required to minimize the hypervisor kernel modifications.

Hyperwall [298] is one of the hypervisor security frameworks which ensures the integrity and confidentiality of the tenant VMs. It helps the system to protect the VMs from untrusted or compromised hypervisor. A small changes is required in the processor design and memory management unit. DMA controls the access to memory regions to ensure the confidentiality in the tenant data. These security policies are integrated with the VMs at the time of booting. A security table is proposed for ensuring the integrity and confidentiality. The access rights of VMs, VMM, and DMA, are maintained in that table which grants or deny access to the memory pages. Here, each microprocessor chip is assigned a key pair. The architecture does not support the detection of hardware attacks.

Hypercoffer is another VMM introspection based framework which helps in maintaining the privacy and integrity of personal and sensitive data of tenants as shown in Figure 8.8 [216]. It also ensures the security to off-chip data from hardware threats. The existing processor is integrated with the encryption and integrity check features. The used address independent seed encryption to protect the memory of VM and Bonsai Merkle tree algorithm for maintaining integrity. The hash values and counters are indexed by guest addresses. Since, it is very difficult to enforce the security policies at the processor level. Therefore, their framework employed software-based solution by adding a security layer between VM and VMM, called as VMShin. The hardware support is provided to VMShin to control the data transfer between guest machine and VMM. The VMShin also helps in minimizing the changes in the hardware design. It also provides the prevention against the attacks targeted from the compromised hypervisor and secure other VMs from attacks. The system is validated on QEMU-emulator and Xen and provides the performance overhead of 0.3–6.8%.

8.3.4 Revisiting the VMM design

There are some security solutions which are based on modifying the VMM design. Some of such approaches are explained below:

NoHype [246] is one of the security architectures which is based on a completely different concept. It provides a different design architecture for virtualization which eliminates the need of hypervisor. It takes the benefit of the existing hardware virtualization and multi-core features of CPU and supports the multi-tenancy. The interaction between guest VM and hardware does not require the involvement of VMM. The management components of VMM are retained so that the administrative functions resuming a VM, stopping or starting VM etc, can be performed easily. In this framework, each VM is allotted with one processor core. The physical memory is partitioned by using the hardware support. The page table mechanism and multicore memory controller (MMC) is used. For each of the CPU core, MMC is configured. This is how the impact of VMs on one-another is minimized. In this framework, it is also assumed that the physical devices are aware about the

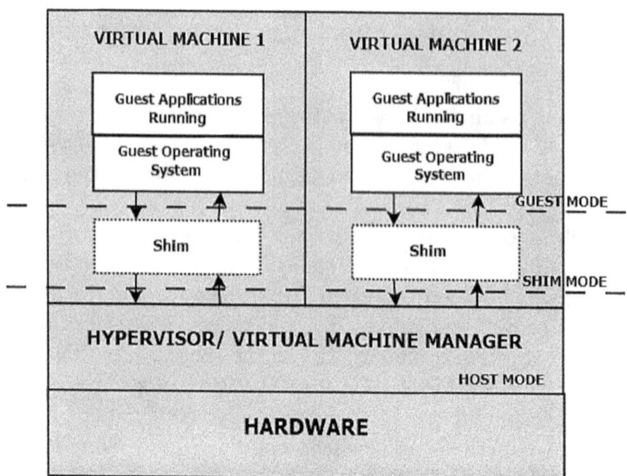

FIGURE 8.8: Conceptual diagram for HyperCoffer.

virtualization. The one physical device can be represented as n virtual physical devices. The design architecture is helpful as a prevention mechanism.

Nova [299] is light weighted micro-VMM kernel-based security framework and uses the Intel-VT support. It minimizes the TCB, thereby removing the attack surface vectors of VMM. Microkernel provides the key abstraction at very basic thread level. If an attacker bypasses the user-level security, it will be very difficult to compromise the kernel code integrity. The full virtualization features are adopted here as well having various components with different access policies. The micro-kernel is executed in the kernel-mode whereas the other components such as root partition manager, device drivers and VMM are executed in the user mode. Various services are provided in this framework such as resource provisioning, interrupt control and management, communication. The performance overhead of the system is reduced to 1–3%. This is applicable to modern commodity hypervisors only.

HyperLock [300] proposed a KVM-based security framework in which a secure separate address space is created for running the VMM kernel. It also provides the prevention of various attacks from the compromised hypervisor. The runtime isolation of VMM and VMM-shadowing are the main key components, provided by this framework. The KVM VMM is loaded and executed in an isolated environment. Any of the guest machine code is not executed in the same space. Jump vectors or trampoline codes are employed to transfer control to host once any guest code arises for execution. KVM can also execute limited instructions. Any attempt to compromise the integrity of the trampoline code is detected by the HyperLock. To execute the trampoline code, the hypervisor invokes the software interrupts. VMM-shadowing allows the compromised KVM to access only one guest which is implemented by some memory scheme. Each guest works on the shadow of the VMM and executes in an isolated environment. The performance overhead is high as system runs various copies of hypervisor. Its challenging to maintain the consistency as well.

8.3.5 VM-assisted hypervisor introspection

The techniques which are discussed above, depends on features of hardware to secure the hypervisor and mainly focused on ensuring the integrity of non-control data (e.g., function pointers). Hyperverify [247] is one of the introspection-assisted hypervisor security frameworks, which monitors the non-control data such as decision making data, configuration data of VMM. The framework introspects the VMM memory from a privileged VM. The VM is trusted and can only be controlled by cloud admin. VM has restricted access than VMM and it is difficult to access the physical memory of VMM from VM. Hence, the direct memory is accessed by DMA by memory access component. The analysis library component tracks the physical address location of VMM. The security policies and privileged data, etc., are analyzed. The contents in the extracted addresses are fetched and matched with some baseline values. The present architecture is very flexible and new modules can easily be integrated. It results into performance loss from 3% to 4%. If the privileged VM is compromised then again it raises a question on security analysis components.

8.4 Conclusion

Cloud computing is prone to various attacking threats including traditional security threats and virualization-specific security threats. Traditional security tools may fail to detect various virtualization specific threats such as VM Escape, cross VM side channel, VMM hyperjacking, etc. In fact, the modern malware are advanced enough and are aware about virtualization. They may not reveal their actual behavior inside the same machine where the security tools are running. It leads toward incorporating the introspection based approaches for detecting the more advanced threats which are occurring in cloud. The key principal behind the VMI is to analyze the VM code or data from outside without getting noticed. However, VMI assumes that VMM is secure. Hence, hypervisor introspection is introduced which performs the analysis of hypervisor code or data structure to ensure the VMM security. Various techniques with details of working mechanisms have been discussed in detail.

8.5 Questions

Fill in the blanks

1. Mark the correct option. The some of the good VMI approaches are

   i LiveWire
   ii Maitland
   iii VAED
   iv All of above

2. Some of the popular introspection based system call tracer tools, are

   i DRAKVUF
   ii Nitro
   iii Ether
   iv All of these

3. Mark the incorrect statement

   i VM state information is very helpful in extracting the useful VM data for analysis.
   ii VM hook based approaches employ an In-VM agent (called hooks) which is injected inside the guest OS of monitored VM.
   iii Hypercalls are also a hardware trap from VM layer to the hardware-layer.
   iv In kernel debugging mechanisms, the kernel data is analyzed by extracting the important kernel logs.

4. Mark the correct statement in the followings

   i Hypervisor introspection mechanisms provide advanced security approaches, capable to detect hypervisor attacks sniffing and network traffic analysis.
   ii VM Introspection approaches are designed to analyze the VM code from an isolated region outside monitored VM.
   iii The VMI libraries/tools help in gaining the VM information and converting it into semantically meaningful form.
   iv All are correct

5. Mark the correct statement. RansomSpector is

   i a VMI-based technique for dealing with crypto-ransomware that runs in the operating system.
   ii a ransomeware attack launching tool
   iii a tool to attack random access memory
   iv None of these

Short-Answer Questions

1. Define introspection. Describe Livewire architecture in detail with suitable diagram.
2. Draw VAED cloud security architecture. Explain its working briefly.
3. How introspection approaches are different than traditional security approaches? Draw a comparison table for VMI techniques/tools.

Long-Answer Questions

1. Draw and explain the taxonomy of hypervisor introspection based approaches.
2. What are the advantages and disadvantages of introspection based approaches?

Chapter 9

Container Security

9.1 Introduction

The most recent decade saw a sensational increment in the turn of events and utilization of virtualization techniques. Henceforth, the interest for a secure and efficient virtualization arrangement (that can give versatile, convenient, thick, and secure user environment) has likewise expanded. An enormous number of such techniques were developed which can be divided into two significant classes, i.e. hypervisor-based and container-based virtualization [301], as shown in Figure 9.1. The last one is developed as an option to virtual machines (VMs). It gives progressively efficient and lightweight environment for users. Additionally, less start-up time is taken by containers which make them a better alternated choice over VMs.

In addition, a container takes less start-up time in 50 milliseconds, while a VM starts in 30–40 seconds. Hence, virtualization using container technology is becoming popular day by day. Containers use the same Kernel as a substitute for receiving a custom copy for each application. In the last few years, the container platform rate has been rising. In 2016, the container market was estimated around 762 million dollars which was anticipated to reach 2.7 billion dollars in 2020 [302]. There exists numerous container advances accessible like Linux-V server, LXCs, Open VZ, with Docker being the newest and the most prevailing candidate.

Docker is a light-weighted container technology with the potential "to construct, run applications and export" [303]. Docker is almost recent and more managing applicant since it appears not to have acquired previous container technologies including the latest capabilities. Docker has several advantages such as speed, fast delivery, density, flexibility. However, there are still a few disadvantages. The absence of total virtualization (as it relies upon the Linux kernel that is given by the nearby host) becomes the major bottleneck while adopting containers. Docker's security is the most essential and vital concern when it comes to adoption of Docker images or working on containerized applications.

In case of VMs, applications can just speak with the kernel of VM and no correspondence exists with the host part. Hence in order to assault the host kernel, the assailant first necessity is to sidestep the hypervisor and then kernel of the virtual machine. However, in the container-based techniques, it isn't the situation. For example, in Docker, applications can legitimately get into and speak with the host kernel, allowing the assailant to attack the host portion straightforwardly [304]. This is one of the key reasons that raise security worries about Docker.

Security is among the predominant challenging situation whenever complex services are primarily inside a multi-tenant organization in virtual environments.

DOI: 10.1201/9781003004486-9

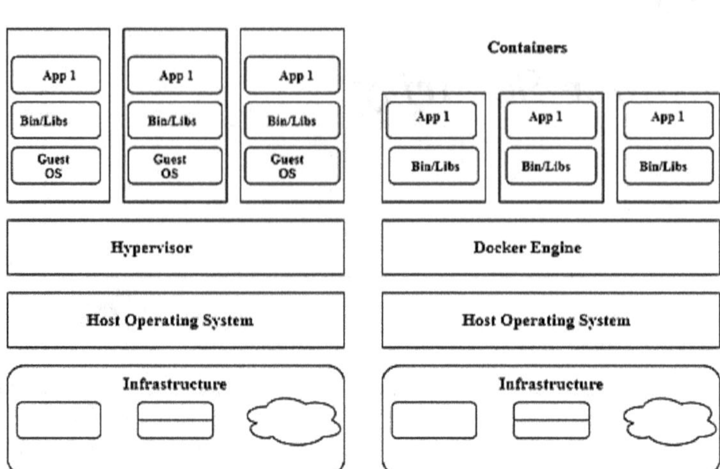

FIGURE 9.1: Virtual machine vs container.

Hypervisor-based techniques are termed to be more secure as a contrast to container-based techniques as former method supports the additional layer of confinement between the application or host. The application running in a virtual environment (VM) can communicate only with VMM/hypervisor and not with the host. Hence, the attacker is mainly responsible for circumventing the hypervisor kernel and then targeting host kernel in a series of process. However, applications can interact with host kernels with all container technologies, and thus allowing an intruder to precisely target a server. An attacker could save a massive number of attempts while partitioning inside the targets host which creates a major security problem with Docker-container.

An intruder can launch various attacks such Denial of Service (DoS), Distributed DDoS, SQL injection and privilege escalation attacks, etc. in containers. Although there exists kernel protection systems which enhance the protection of a Linux host machine such as Linux Security component (LSM) and Linux capabilities. The LSM provides a system to support restricted security models in the Linux kernel. With security being such a significant issue, there are numerous assaults conceivable in Docker on its different segments like containers, applications, hosts, and Docker engine. Docker's poisoned images are also serious threats to entire container sever. Denial of Service (DoS) and Distributed Denial of Service (DDoS) are the common ones as these attacks can be easily launched due to the availability of multiple attacking tools and can cause huge financial losses to the organizations [305]. Moreover, the weak isolation provided by containers raises very high possibility of the attacks taking place in the Docker environment. Let's first discuss the threat model and various attacks in detail in the following sections.

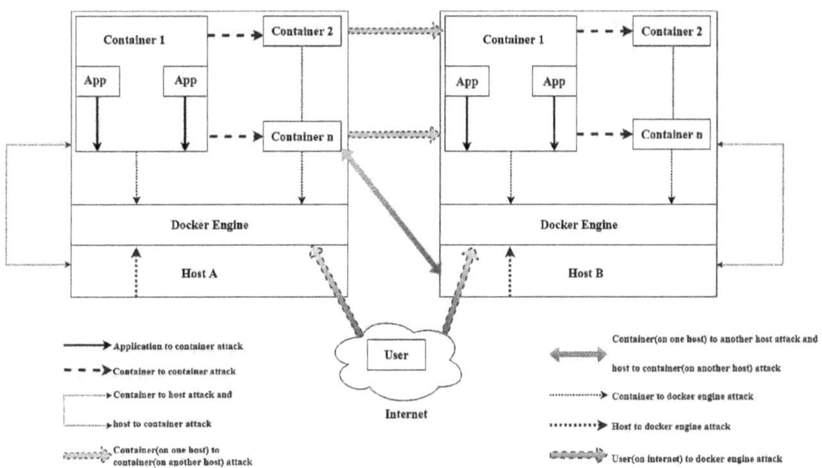

FIGURE 9.2: Threat model.

9.2 Threat Model in Containerized Environment

In this section, a threat model is explained which is the expanded version of the threat model, proposed by Sultan et al. [302]. The threat model considers the multi-node set up while explaining various attack surfaces. It covers all the possible attack scenarios in multiple hosts, as shown in Figure 9.2. Let's first understand the terms such as semi-trustable and untrustable adversary which are being used during the discussion below. An untrustable adversary is a type of active adversary that can break any sort of protocol in order to steal data. Whereas a semi-trustable adversary being a passive entity will not break any protocol, but it can help in the stealing of data. Trustable platform refers to the platform which is safe from attacks. Let's now understand various attacking scenarios.

Case 1: Targeting a Container from Application

The assumption that is being made in this case is, applications cannot bypass the control policies if they are already set. In addition, certain applications might require the root privileges in order to run. Now if any application will get access to the control manager then it can cause damage to the host as well as the containers running inside the host. So, the containers should be anyhow protected from such applications.

Case 2: Targeting a Container from Co-located Container

The assumption, being made in this case is, either one or more than one containers and applications running inside those containers are semi-trustable or untrustable. The aim is to protect all the containers from every other container which is present inside the host. Now, if there is a proper isolation provided, each container will not know anything about the other containers running in the same host. It will then become difficult to target the co-located containers.

Case 3: Targeting a Host from Container

The assumption, considered in this case is, at least one container present in the host and the applications running in the trustable containers are either semi-trustable or untrustable. In this case as well, the weak isolation can cause attackers to bypass the memory region allocated to containers and target host kernel. It should be ensured that the containers cannot make any compromise with the privacy, probity, and accessibility of host constituents.

Case 4: Targeting a Container from Host

The assumption, being made in this case is, the host can be semi-trustable or untrustable but the containers running on the top of it, are trustable. The security of containers can be breached by untrustable host. The containers running inside such hosts can easily get compromised through multiple attacks (include both active and passive attacks). Active attacks can cause more damage to the containers in comparison to passive attacks.

Case 5: Targeting a Container from Another Container Running in Different Host

The assumption made in this case is, the same as that of case II. The aim is to protect the containers not only from the containers present on the same host but also from the containers which are present on another host. Now if the containers will behave like virtual machines, i.e. the containers will not be aware of the existence of other containers until it is required then it will be a perfect solution for the attacks that will take place under this case.

Case 6: Targeting a Host from Container Running in Different Host

In order to tackle the attacks that can take place in this case, no container should try to compromise with the CIA properties of all the other constituents present in the host. This can happen only when all the containers try to behave like virtual machines.

Case 7: Targeting a Container (Running in One Host) from Another Host

The containers present on one host needs to be protected from the containers present on another host. This can happen only when the containers present on different hosts don't know anything about each other and communicate only when it is required.

Case 8: Attack on Docker Engine from Container

As the containers having the required access rights can easily cause damage to the Docker engine so, in order to prevent this from happening, access rights should be provided to the trustable users only.

Case 9: Targeting Docker Engine from Host

The host has all the privileges so it can misuse its privileges and can vandalize the Docker engine. To avoid this, all such scenarios need to be eliminated where the misuse of privileges can happen.

Case 10: Targeting Docker Engine from User (Present on the Internet)

This case can take place in a cloud-based environment. In such a scenario, the user can attack the Docker engine by making use of the Cloud Container Attack Tool (CCAT) [306].

9.2.1 Attacks in containers

There are various attacks which are happening in container environment. These attacks can target containers, applications, container engine and host, etc. Let us discuss some of the key attacks.

ARP Spoofing and MAC Flooding Attacks

Connectivity between the host machine and container is provided by a virtual ethernet bridge. The creation of a new container leads to the establishment of a new virtual Ethernet interface that will be connected with the bridge. As no filtering is performed during the forwarding of all the incoming packets to the desirable interfaces, hence the model described above can become an easy target for the attacks like ARP spoofing and MAC flooding [304].

Attack Using CCAT

A tool known as CCAT was developed to check the safety arrangements of a cloud-based environment but later on, the same tool was used by the attackers to exploit the environment of Amazon Web Services (AWS) by causing damage to the AWS credentials [306]. Containers can be deployed on AWS by making use of Amazon

Elastic Container Service (ECS) which is a highly flexible service for container management having very high performance.

Attack on Unnecessary Services

The programs that pay attention and give replies to the incoming and outgoing traffic are called services. FTP servers, proxy servers, web servers, and many other servers can be directly accessed using services. But if more services are running inside our system then there will be more chances for attackers to break into them and gain control of the system.

Container Escape Attack

Root privileges are required by the attacker in order to gain access to the required resources. Now if this attack gets launched successfully, the attacker will not only acquire root privileges of the host (in order to attack the required resources) but will also become able to deploy a malicious container inside our environment.

Cryptojacking

In this attack, the computational power of the infected machines will be stolen by the attackers. This will be done by robbing the CPUs with the help of malware. The purpose of this attack is to mine for virtual coins (like Monero (XMR) and Ethereum (ETH)) that will be later sent to the wallets owned by the attackers [307].

Denial of Service (DoS) Attack

This attack can make a system or network unavailable to legitimate users by continuously sending data packets or requests. The number of these packets or requests is more than the capacity of the system or network so the system or network will eventually get down. This attack is mainly of two types: (1) disability and (2) resource exhaustion. The situation of disability is already defined in the definition. Whereas in resource exhaustion, resources like memory and CPU are overly consumed by the attacker and due to this, these resources will not be available to the legitimate users.

Kernel Exploit

Under this attack, the applications present inside the containers can get exploit. As the same kernel is shared among all the containers hence if any application gets attacked and become successful in obtaining the root privileges associated with the kernel, then not only the running containers but also the entire host will be at stake.

Malicious Code Injection

An attacker can inject a malicious program into a container thus making the container infected and this infected container can then attack the host, hence this problem became a huge security issue in containers. As per a newly found vulnerability [28], root privileges can be obtained by an infected container by overwriting the host runc library. This entire thing can be initiated by the attacker just by placing a container infected inside the system which is again not a very tedious task to perform.

Man-in-the-Middle (MitM) Attack

In this attack, a communication between two legitimate users is monitored by an attacker. The attacker then tries to modify and rob the confidential information getting exchanged between the users.

Malware

If a victim opens a malicious web page, the Docker Build command (which is used to create a container) can then be executed by the attacker and REST API will help in the execution of this command. Once the attacker bypasses same-origin policy by making use of Host Rebinding technique, it will obtain the access to reach host machine. The malware will then be able to run on the user's machine or the attacker can make the Docker images poisoned so that the malware will spread as soon as we launch Docker.

Poisoned Images

The image getting downloaded by a user from Docker hub can be poisoned and can exploit not only the applications and containers but the entire host platform as well. Using insecure and outdated software or injecting malicious software into the system can also lead to the problem of poisoned images.

Privilege Escalation

The attacker tries to gain the root privileges of the kernel through this attack. There are two methods through which this attack can be performed: (1) file modification (2) memory modification. Using file modification, the attacker tries to modify the privileged files in order to alter the file attributes or to change the super user's password. Whereas in memory modification, the attacker tries to change the control flow by overwriting certain data structures in the memory.

Tampering

If the mounting of confidential directories on the host operating system is performed by a container, then the same container can make changes in the files which are present in those directories. The stability and safety of the host as well as all the containers running inside it can get affected by these changes. Hence the configurations which are insecure for the containers can lead to the possibility of tampering inside the host.

9.3 Defense Mechanisms

Some key research works have been taken into consideration that deals with Docker and its security aspects. A brief summary of each of the research work is given in the section.

Sultan et al. [302] made a comparison between containers and VMs. A threat model having four use cases was also proposed for the containers. Among the four, only one use case makes use of the hardware-based solutions, whereas the other three use cases make use of software-based solutions. Linux kernel features and Linux Security Modules are discussed under software-based solutions, whereas Intel SGX and virtual Trusted Platform Modules are discussed under hardware-based solutions. Lastly, some open issues and the possibilities for future work are presented.

Rad et al. [303] provided the pros and cons of Docker along with a short description of the four constituents of Docker, namely containers, images, Docker client and server, and Docker registries. A comparison of Docker with other virtualization techniques like XEN, LXCs, and KVM has also been made.

Scheepers [308] did a comparison between Linux Containers (LXC) and XEN and provided the advantages and disadvantages for both. It was concluded that XEN is better when it comes to equal distribution of resources, whereas LXC is better in the terms of resource utilization as it wastes fewer resources but the time taken to complete the tasks is more in the case of LXC. Felter et al. [309] compared KVM and Docker and found out that if we talk about the resources getting waste, then there is no major difference between these two as they both waste resources equally. Seo et al. [310] also made a comparison between KVM and Docker in terms of performance and came to a conclusion that Docker has a better performance than KVM.

Tien et al. [311] proposed a tool known as KubAnomaly that can be used to detect anomalies in Kubernetes, which is a cloud container orchestration platform. System logs were monitored in order to propose a feature extraction method. Three different datasets (publicly available, private, and data of the experiments conducted in the real world) were used in order to evaluate the performance of the proposed system. The potency of the model that has been proposed was illustrated by making a comparison between its accuracy and accuracy of other machine learning algorithms and the accuracy was found to be 96%. The proposed model successfully identified four real attacks that were launched by attackers in September 2018.

Chelladhurai et al. [304] presented various threats to Docker under which attacks like ARP spoofing, MAC spoofing, and DoS are discussed in brief. The work which is carried out to address the security of Docker has also been discussed. A solution was proposed to prevent Docker from DoS attacks. Under this solution, the main focus was to control the memory limit. This solution has been classified in three test cases; in the first test case, memory limit has been assigned in the Docker run command; in the second test case, it has been assigned in the default value of the source code, whereas in the last test case a default value is to be provided to the memory limit so that any DoS attack will not result in the system crash.

Abed et al. [312] described system overview using Linux kernel. They employed the system call frequency-based Bag of system call (BoSC) approach, which is a feature extraction approach. They performed malicious attack using SQL map on the container to create malicious dataset by targeting the container-organized Mysql database. The authors used Linux *strace* to trace all system calls. The approach has been validated in their self-generated dataset and results seem to be promising.

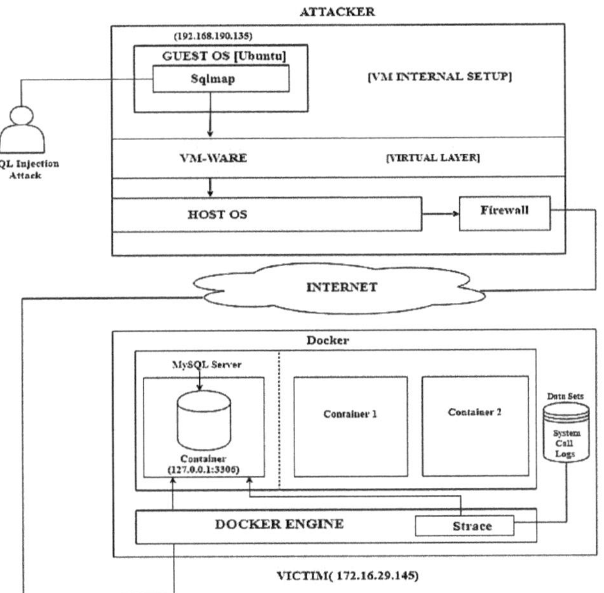

FIGURE 9.3: Testbed environment for SQL injection attack.

9.4 Case Study on SQL Injection Attack in Containers

Web applications are prone to lots to application layer attacks. SQL injection attack is one such attack in which malicious scripts are inserted into strings which later brought to the processing or implementation of database backend. The attack is carried out from attacker machine (Kali Linux) to victim machine (Linux container) connected on the same network as shown in Figure 9.3. The below tools have been used in this case study for setting up environment, launching the attack and extracting the attacking logs.

- Sqlmap: It is a free and open source scanning tool which automatically detects and exploits SQL injection bugs or takes along web server. This comes with an advanced remote system, numerous specialty attributes only for overall legitimate user, or a wide range with controller from server authentication, query sniffing from its data file, attempting to access a fundamental system files, or implementing command line.
- Strace: It would be a testing, logging, or instruction Linux kernel tool. This can be used for controlling or manipulating connections among programs as well as the Linux, including process state changes, signal distribution or system calls.
- Mysql-server: Mysql Server is a tool for the management of databases. A database system arranges details under one and many database table that can

connect data structures to one another; such relationships help in facilitating a database.

The attacker machine was running Kali Linux having Sqlmap pre-installed. The victim machine (Ubuntu 18.04.3 LTS, 16 GB, 1 TB, Core i5, Docker ver. 19.03.05 (community ed.), Mysql 5.6 docker image), was running a Linux container, a primary target of the attacker. Whenever container performs some activity, the system uses the strace tool to trace all system calls to capture the run time behavior of the Mysql-server server. We have installed Docker in the machine by following [313] commands. The step-wise procedure to install required tools and attack launching is explained below.

9.4.1 Part-A-test bed set up

Lets us now explain the step by step procedure for setting up test bed.

Step 1

The docker pull command in docker is used to pull image of your container from official docker storage pool. The command output is given in Figure 9.4.
$ sudo docker pull mysql/mysql-server

```
geu@geu-ThinkCentre-M82:~$ sudo docker pull mysql/mysql-server:latest
latest: Pulling from mysql/mysql-server
c7127dfa6d78: Pull complete
530b30ab10d9: Pull complete
```

FIGURE 9.4: Step 1: Pulling docker image.

Step 2

Verifying mysql image. You can list the downloaded Docker images using the command, given in Figure 9.5.
$ sudo docker images

```
geu@geu-ThinkCentre-M82:~$ sudo docker images
REPOSITORY            TAG       IMAGE ID       CREATED       SIZE
mysql/mysql-server    latest    a7a39f15d42d   2 days ago    381MB
```

FIGURE 9.5: Step 2: Verifying mysql image.

Step 3

Start running a docker container from Mysql image. You can start mysql-server with the docker run command as below (refer Figure 9.6):
$ docker run – name=mysql11 –d mysql/mysql-server

Here docker run command is used to run a container.
-name is used to give name to a container if not used a random name will be

Try Ubuntu without installing
Install Ubuntu
Check disc for defects
Test memory
Boot from first hard disk

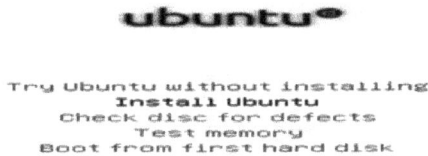

FIGURE 9.6: Step 3: Running docker container.

assigned to the container. -d switch is used when we want to run the container in the background.

Step 4

Stop and remove the container and run again as below (refer Figure 9.7):
$ Docker stop mysql1 & docker rm mysql1

```
geu@geu-ThinkCentre-M82:~$ sudo docker stop mysql1 && sudo docker rm mysql1
mysql1
mysql1
```

FIGURE 9.7: Step 4: Stop and remove the docker container.

Step 5

Binding port for external connection (refer Figure 9.8):
If you only want to access it locally, it's enough, but if you want to access your mysql server remotely, you need mapping the port 3306 of the container to port 3306 of the host machine. Also, you need to set the environment variable MYSQL_ROOT_HOST with wildcards to allow root connections from other hosts:
$ sudo docker run – name=mysql1 –e MYSQL_ROOT_HOST=% -p 3306 –d mysql/mysql-server Here -p is used to give port number that we bind to the mysql server.

```
geu@geu-ThinkCentre-M82:~$ sudo docker run --name=mysql -e MYSQL_ROOT_HOST=% -p 3306:3306 -d mysql/mysql-server
13b26f87388d2f588d886119e8d9df405c1e81b9585c781689cd7f9c08923885
```

FIGURE 9.8: Step 5: Run the container again with port binding mysql server.

Step 6

Logging into mysql part 1 (refer Figure 9.9):
Since it's already running, you can connect to it with the docker exec command
$ sudo docker logs mysql1 2>1

```
geu@geu-ThinkCentre-M82:~$ sudo docker logs mysql 2>&1 | grep GENERATED
geu@geu-ThinkCentre-M82:~$ sudo docker logs mysql 2>&1
[Entrypoint] MySQL Docker Image 8.0.19-1.1.15
[Entrypoint] No password option specified for new database.
```

FIGURE 9.9: Step 6: Connect to Mysql.

Step 7

Logging into mysql server (refer Figure 9.10)
$ sudo docker logs mysql1

```
geu@geu-ThinkCentre-M82:~$ sudo docker logs mysql 2>&1 | grep GENERATED
geu@geu-ThinkCentre-M82:~$ sudo docker logs mysql 2>&1
[Entrypoint] MySQL Docker Image 8.0.19-1.1.15
[Entrypoint] No password option specified for new database.
```

FIGURE 9.10: Step 7: Logging to server.

Step 8

Install strace: strace is a system call tracer (refer Figure 9.11), i.e. a debugging tool which prints out a trace of all the system calls made by a another process/program.
$ sudo apt-get install strace –y

```
geu@geu-ThinkCentre-M82:~$ sudo apt-get install strace -y
Reading package lists... Done
Building dependency tree
Reading state information... Done
```

FIGURE 9.11: Step 8: Installation of strace.

Step 9

Altering user png as shown in Figure 9.12.
ALTER USER 'username'@'localhost'IDENTIFIED BY 'Password'

```
mysql> ALTER USER 'root'@'localhost' IDENTIFIED BY '<malware>';
Query OK, 0 rows affected (0.14 sec)

mysql> SHOW DATABASES;
+--------------------+
| Database           |
+--------------------+
| information_schema |
| mysql              |
| performance_schema |
| sys                |
+--------------------+
4 rows in set (0.01 sec)

mysql> ■
```

FIGURE 9.12: Step 9: Alerting a user.

Step 10

Creation of database in mysql-server (refer Figure 9.13)
$ sudo docker exec -it mysql1 mysql –u root -p

```
geu@geu-ThinkCentre-M82:~$ sudo docker exec -it mysql1 mysql -u root -p
Enter password:
Welcome to the MySQL monitor.  Commands end with ; or \g.
Your MySQL connection id is 130
Server version: 8.0.19 MySQL Community Server - GPL

Copyright (c) 2000, 2020, Oracle and/or its affiliates. All rights reserved.

Oracle is a registered trademark of Oracle Corporation and/or its
affiliates. Other names may be trademarks of their respective
owners.

Type 'help;' or '\h' for help. Type '\c' to clear the current input statement.

mysql> CREATE DATABASE malware;
Query OK, 1 row affected (0.11 sec)

mysql> USE malware;
Database changed
mysql> CREATE TABLE malinfo ( id smallint unsigned not null auto_increment, name
 varchar(20) not null, constraint pk_example primary key (id) );
Query OK, 0 rows affected (0.65 sec)
```

FIGURE 9.13: Step 10: Creation of Mysql database.

Step 11

Pulling phpMyAdmin/phpMyAdmin image from docker repository (refer Figure 9.14).
$ sudo docker pull phpmyadmin/phpmyadmin

```
geu@geu-ThinkCentre-M82:~$ sudo docker pull phpmyadmin/phpmyadmin
[sudo] password for geu:
Using default tag: latest
latest: Pulling from phpmyadmin/phpmyadmin
Digest: sha256:2eccbe375bffb5ddd9cf63a4544c9a48a78b1a8138b728cb0af1e0d4937a340a
Status: Image is up to date for phpmyadmin/phpmyadmin:latest
docker.io/phpmyadmin/phpmyadmin:latest
```

FIGURE 9.14: Step 11: Pulling the phpMyAdmin image.

Step 12

Making a container for Php image to host mysql server locally and binding mysql-server with it (refer Figure 9.15).
$ sudo docker run –name myadmin –d –link mysql1:db –p 8080:80 phpmyadmin/phpmyadmin

```
rs@rs-virtual-machine:~$ sudo docker run --name myadmin -d --link mysql1:db -p 8080:80 php
myadmin/phpmyadmin
42959274246edbe2b61ec2099a67f8e8e8a2dc24094c0e539c286159927a8809
rs@rs-virtual-machine:~$ sudo docker ps -a
CONTAINER ID        IMAGE                    COMMAND                  CREATED             S
TATUS                 PORTS                              NAMES
42959274246e        phpmyadmin/phpmyadmin    "/docker-entrypoint.…"   6 seconds ago       U
p 4 seconds             0.0.0.0:8080->80/tcp               myadmin
ef17bd354eb2        mysql/mysql-server       "/entrypoint.sh mysq…"   2 hours ago         U
p 2 hours (healthy)     0.0.0.0:3306->3306/tcp, 33060/tcp   mysql1
```

FIGURE 9.15: Step 12: Running container from Php image.

Step 13

As there will be a connect error for this when logging in, so to correct it change the hashing password with your password as shown in Figure 9.16.

```
mysql> ALTER USER 'root'@'localhost' IDENTIFIED WITH caching_sha2_password BY '<password>
';
Query OK, 0 rows affected (0.07 sec)

mysql> ALTER USER 'root'@'%' IDENTIFIED WITH caching_sha2_password BY '<password>';
Query OK, 0 rows affected (0.01 sec)
```

FIGURE 9.16: Step 13: Password change.

9.4.2 PART B: Attacking launching and malicious logs extraction

Let us now understand the step-by-step procedure for attacking the machine and extracting logs.

Step 14

Now sql queries are injected into the server (refer Figure 9.17)
$ sudo sqlmap –u http://192.168.35.128:8080/db_structure.php?db= mallu –dbs –text-only –level=5 –TECH= "BEUSTQ"
Here -u is used to provide the victim link. And –dbs is used to forcefully reveal the databases. The –text-only switch is used to set sqlmap to provide only text related queries. –level switch configures the sqlmap to try more variations of queries. For different techniques there is a switch paramenter as –TECH= "technique name". For more details, please refer to sqlmap –help.

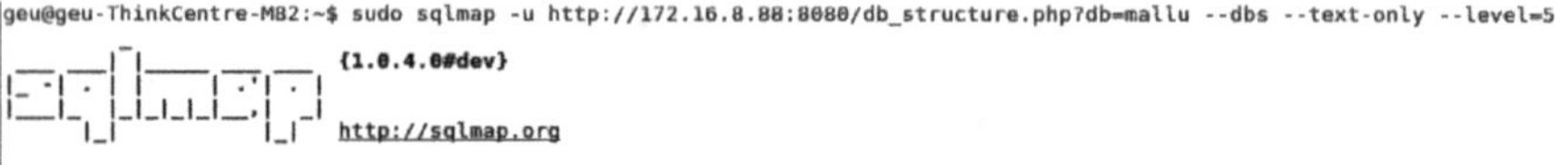

FIGURE 9.17: Step 14: Inject sql queries into server.

Step 15

Logs for sql injection. The output log, generated by below command is shown in Figure 9.18.
$ sudo strace -ff –p 14521 -o geusqlmap.text

A case study is explained above in detail to provide practical knowledge about attacks in containers. We have successfully launched and extracted logs of attacks which can be further analyzed by using intrusion detection techniques.

```
select(0, NULL, NULL, NULL, {tv_sec=0, tv_usec=491}) = 0 (Timeout)
futex(0x56047d5d9280, FUTEX_WAIT_BITSET_PRIVATE|FUTEX_CLOCK_REALTIME, 0, NULL, 0xffffffff) = -1 EAGAIN (Resource temporarily una
select(0, NULL, NULL, NULL, {tv_sec=0, tv_usec=10000}) = 0 (Timeout)
futex(0x56047d5d9280, FUTEX_WAIT_BITSET_PRIVATE|FUTEX_CLOCK_REALTIME, 0, NULL, 0xffffffff) = 0
select(0, NULL, NULL, NULL, {tv_sec=0, tv_usec=10000}) = 0 (Timeout)
socket(AF_INET, SOCK_STREAM, IPPROTO_TCP) = 10
fcntl(10, F_GETFL)                       = 0x2 (flags O_RDWR)
fcntl(10, F_SETFL, O_RDWR|O_NONBLOCK)    = 0
connect(10, {sa_family=AF_INET, sin_port=htons(8080), sin_addr=inet_addr("192.168.35.128")}, 16) = -1 EINPROGRESS (Operation now
poll([{fd=10, events=POLLOUT}], 1, 30000) = 1 ([{fd=10, revents=POLLOUT}])
getsockopt(10, SOL_SOCKET, SO_ERROR, [0], [4]) = 0
select(0, NULL, NULL, NULL, {tv_sec=0, tv_usec=10000}) = 0 (Timeout)
futex(0x56047d5d9280, FUTEX_WAIT_BITSET_PRIVATE|FUTEX_CLOCK_REALTIME, 0, NULL, 0xffffffff) = 0
```

FIGURE 9.18: Step 15: Process logs for sql injection.

9.5 Open Research Challenges for Container Security

There are various open challenge in container security. Some of the key challenges are listed below:

No Publicly Available Dataset

Many older datasets like KDD [157], NSL-KDD'99 [314], and UNSW-NB15 [158] are available. However, they are outdated for the current virtualization scenario. On the basis of several studies [315], it was found out that the older datasets have their own set of limitations which limit their usability. Some of the limitations with widely used datasets are listed below:

- KDDCUP'99 [157] contains a large number of redundant records in the training set which may cause a classifier to be biased toward more frequent record and the presence of a large number of missing records also becomes the reason for the change in data.
- NSLKDD dataset [314] which is the upgraded version of the KDDCUP 99 dataset solves many issues that were present in KDDCUP 99 but it is also not found to provide expected and appropriate results in the modern attacking environment [158].
- The CAIDA datasets [316] did not have a ground truth about the attack instances and the collected data was not processed to produce new features that could improve the differentiation between attack and normal trace.

Though the target addresses of the packets can be changed in the older datasets and eventually can be directed to a test machine while replaying, the real effects of the network trace cannot be reproduced because of the unavailability of the services addressed in the original trace [315].

Limitations with Existing Detection and Mitigation Algorithm

Just like dataset, many algorithms are present but none of them are proposed and implemented by keeping the Docker-based environment in mind. As a result, they may or may not work in the Docker-based environment.

Probity of Docker's Host

There's a big question on Docker's daemon reliability as there's no guarantee whether it is reliable or not. Also, how to guarantee the trustworthiness when Docker host is getting booted? Another major task is to check the trustworthiness of Docker containers and the security of the Docker engine must be ensured.

Container's Security Standards

Further examination for normalization of container organization, correspondence conventions, and evaluation philosophies is required. There's likewise a need to additionally inspect the practicality of consequently moderating the found compartment vulnerabilities by applying the necessary arrangements naturally.

The Usefulness of Risk Assessment Tools

Several analysts demonstrated that 30–90% high-hazard risks can be present in Docker images. The same number of powerlessness evaluation devices are accessible for Docker images, the inquiry here is about the convenience of such tools in the creation condition and if their utilization could obstruct the arrangement procedure or not.

Container Alternatives

Further examination is required on containers, virtual machines, unikernels, and hybrids. Research can be conducted on the various possibilities of hybrid combinations and the benefits that can be achieved through hybrids. This would help recognize the best innovation for various situations.

Malicious Docker Software

Docker images come with a large software packages. If there is any vulnerability present in any software, it may lead to a crucial security risk. The containers can also be exposed to serious security threats if not proper isolation is provided. It may cause data leakage, data modification, or data theft, etc. The docker vulnerabilities should be tracked and handled on time in order to mitigate the threats.

9.6 Conclusion

Since the past two decades, virtualization techniques were around us; however containers are the newest and the most superior candidates among all the existing virtualization techniques. Container's security is the primary and the most crucial concern due to which there is a very high possibility of the attacks. A Docker environment has been considered for study. A threat model and various attacks are explained in detail. Various defensive mechanisms are also discussed following the various research gaps. At the end, a case study is provided to demonstrate how sql injection attack can be performed in Docker machines to give a practical insight to the readers. The study explains textbed environment required, different types of tools such as sql map, strace, strace-docker, Mysqlserver. The attack is carried out from attacker machine to victim container container machine.

9.7 Questions

Fill in the blanks

1. Mark the incorrect option.
   - i A container takes start-up time around 50 seconds
   - ii A VM takes start-up time around 30-40 seconds
   - iii The container present on one host needs to be protected from the containers present in another host.
   - iv The user can attack the Docker engine by making use of cloud container attacking tools.
2. Which of the following attack is not the part of container environment
   - i Cryptojacking
   - ii Poisoned images
   - iii Man-in-the-middle attack
   - iv Hyperjacking attack
3. Mark the incorrect statement
   - i In cryptojacking attack, the computational power of the infected machines will be stolen by the attackers.
   - ii An attacker can inject a malicious program into a container thus making the container infected.
   - iii Docker is a heavy weighted container technology with the potential to construct and run applications.
   - iv Sqlmap is a free and open-source scanning tool which automatically detects and exploits SQL injection bugs.
4. Mark the correct statement in the followings
   - i DDoS attack can make a system or network unavailable to legitimate users by continuously sending data packets or requests.
   - ii The image getting downloaded by a user from Docker hub can be poisoned.
   - iii The attacker tries to gain the root privileges of the kernel through privilege escalation attack.
   - iv All are correct

5. Hypervisor-based techniques are termed to be more secure as a con- trast to container-based techniques. Is this statement true:

   i Yes
   ii No
   iii May be
   iv None of these

Short-Answer Questions

1. Draw and explain threat model in Docker containers.
2. How containers are better than hypervisors?
3. Write at-least five research challenges in container environment.

Long-Answer Questions

1. Draw the basic architecture of VM-versus-Docker. Discuss the advantage and disadvantages of both.
2. List and explain various Docker attacks in detail. Write some attacking tools as well.

Bibliography

[1] R. Buyya, C. Vecchiola, and S. T. Selvi, *Mastering cloud computing: foundations and applications programming.* Newnes, 2013.

[2] Q. Zhang, L. Cheng, and R. Boutaba, "Cloud computing: state-of-the-art and research challenges," *Journal of internet services and applications*, vol. 1, no. 1, pp. 7–18, 2010.

[3] S. I. Bairagi, and A. O. Bang, "Cloud computing: History, architecture, security issues," in *National Conference "CONVERGENCE*, vol. 2015, 2015, p. 28.

[4] P. Mell, T. Grance *et al.*, "The nist definition of cloud computing," Tech. Rep., 2011.

[5] I. Foster, Y. Zhao, I. Raicu, and S. Lu, "Cloud computing and grid computing 360-degree compared," in *Grid Computing Environments Workshop.* IEEE, 2008, pp. 1–10.

[6] N. Phaphoom, N. Oza, X. Wang, and P. Abrahamsson, "Does cloud computing deliver the promised benefits for it industry?" in *Proceedings of the WICSA/ECSA 2012 Companion Volume*, 2012, pp. 45–52.

[7] Openstack, *OpenStack: Open Source Cloud Computing Software*, 2015. [Online]. Available: https://www.openstack.org/software/

[8] T. Benson, A. Akella, A. Shaikh, and S. Sahu, "Cloudnaas: a cloud networking platform for enterprise applications," in *Proceedings of the 2nd ACM Symposium on Cloud Computing*, 2011, pp. 1–13.

[9] D. F. Parkhill, *Challenge of the computer utility.* Addison-Wesley, 1966.

[10] P. E. Ceruzzi, E. Paul, W. Aspray *et al.*, *A history of modern computing.* MIT Press, 2003.

[11] R. P. Goldberg, "Survey of virtual machine research," *Computer*, vol. 7, no. 6, pp. 34–45, 1974.

[12] S. Ma, "A review on cloud computing development," *Journal of Networks*, vol. 7, no. 2, p. 305, 2012.

[13] I. A. T. Hashem, I. Yaqoob, N. B. Anuar, S. Mokhtar, A. Gani, and S. U. Khan, "The rise of "big data" on cloud computing: Review and open research issues," *Information Systems*, vol. 47, pp. 98–115, 2015.

[14] L. Qian, Z. Luo, Y. Du, and L. Guo, "Cloud computing: An overview," in *IEEE International Conference on Cloud Computing.* Springer, 2009, pp. 626–631.

[15] J. Voas, and J. Zhang, "Cloud computing: New wine or just a new bottle?" *IT Professional*, vol. 11, no. 2, pp. 15–17, 2009.

[16] M. S. Elliott, and K. L. Kraemer, *Computerization movements and technology diffusion: From mainframes to ubiquitous computing*. Information Today, Inc., 2008.

[17] R. Buyya *et al.*, *High performance cluster computing: Architectures and systems (volume 1)*. Prentice Hall, Upper SaddleRiver, NJ, vol. 1, p. 999, 1999.

[18] I. Foster, C. Kesselman, and S. Tuecke, "The anatomy of the grid: Enabling scalable virtual organizations," *The International Journal of High Performance Computing Applications*, vol. 15, no. 3, pp. 200–222, 2001.

[19] Z. Juhász, P. Kacsuk, and D. Kranzlmuller, *Distributed and parallel systems: Cluster and grid computing*. Springer Science & Business Media, 2004, vol. 777.

[20] B. Clark, T. Deshane, E. M. Dow, S. Evanchik, M. Finlayson, J. Herne, and J. N. Matthews, "Xen and the art of repeated research." in *USENIX Annual Technical Conference, FREENIX Track*, 2004, pp. 135–144.

[21] T. O'Reilly, "Web 2.0: compact definition," 2005.

[22] M. P. Papazoglou and D. Georgakopoulos, "Service-oriented computing," *Communications of the ACM*, vol. 46, no. 10, pp. 25–28, 2003.

[23] M. J. Buco, R. N. Chang, L. Z. Luan, C. Ward, J. L. Wolf, and P. S. Yu, "Utility computing SLA management based upon business objectives," *IBM Systems Journal*, vol. 43, no. 1, pp. 159–178, 2004.

[24] F. Liu, J. Tong, J. Mao, R. Bohn, J. Messina, L. Badger, and D. Leaf, "Nist cloud computing reference architecture," *NIST special publication*, vol. 500, no. 2011, p. 292, 2011.

[25] M. J. Kavis, *Architecting the cloud: Design decisions for cloud computing service models (SaaS, PaaS, and IaaS)*. John Wiley & Sons, 2014.

[26] M. Cusumano, "Cloud computing and SaaS as new computing platforms," *Communications of the ACM*, vol. 53, no. 4, pp. 27–29, 2010.

[27] C. Pahl, "Containerization and the PaaS cloud," *IEEE Cloud Computing*, vol. 2, no. 3, pp. 24–31, 2015.

[28] S. Bhardwaj, L. Jain, and S. Jain, "Cloud computing: A study of infrastructure as a service (IaaS)," *International Journal of engineering and information Technology*, vol. 2, no. 1, pp. 60–63, 2010.

[29] L. Savu, "Cloud computing: Deployment models, delivery models, risks and research challenges," in *2011 International Conference on Computer and Management (CAMAN)*. IEEE, 2011, pp. 1–4.

[30] C. C. Rao, M. Leelarani, and Y. R. Kumar, "Cloud: computing services and deployment models," *International Journal of Engineering and computer science*, vol. 2, no. 12, pp. 3389–3392, 2013.

[31] M. A. Bamiah, and S. N. Brohi, "Exploring the cloud deployment and service delivery models," *International Journal of Research and Reviews in Information Sciences*, vol. 1, no. 3, pp. 77–80, 2011.

[32] S. Goyal, "Public vs private vs hybrid vs community-cloud computing: a critical review," *International Journal of Computer Network and Information Security*, vol. 6, no. 3, p. 20, 2014.

[33] B. Wilder, *Cloud architecture patterns: using microsoft azure.* O'Reilly Media, Inc., 2012.

[34] S. Krishnan, and J. L. U. Gonzalez, *Building your next big thing with google cloud platform: A guide for developers and enterprise architects.* Springer, 2015.

[35] A. Kochut, Y. Deng, M. R. Head, J. Munson, A. Sailer, H. Shaikh, C. Tang, A. Amies, M. Beaton, D. Geiss *et al.*, "Evolution of the ibm cloud: Enabling an enterprise cloud services ecosystem," *IBM Journal of Research and Development*, vol. 55, no. 6, pp. 1–7, 2011.

[36] J. Smith, and A. C. Team, *Adobe creative cloud design tools digital classroom.* John Wiley & Sons, 2014.

[37] P. Paul, P. Aithal, R. Saavedra M, P. S. B. Aremu, and P. Baby, "Cloud service providers: An analysis of some emerging organizations and industries," *International Journal of Applied Engineering and Management Letters (IJAEML)*, vol. 4, no. 1, pp. 172–183, 2020.

[38] J. Nieh, and O. C. Leonard, "Examining vmware," *Dr. Dobbâ€™s Journal*, vol. 25, no. 8, p. 70, 2000.

[39] M. Mao and M. Humphrey, "A performance study on the vm startup time in the cloud," in *2012 IEEE Fifth International Conference on Cloud Computing.* IEEE, 2012, pp. 423–430.

[40] S. Wu, L. Deng, H. Jin, X. Shi, Y. Zhao, W. Gao, J. Zhang, and J. Peng, "Virtual machine management based on agent service," in *2010 International Conference on Parallel and Distributed Computing, Applications and Technologies.* IEEE, 2010, pp. 199–204.

[41] C. Clark, K. Fraser, S. Hand, J. G. Hansen, E. Jul, C. Limpach, I. Pratt, and A. Warfield, "Live migration of virtual machines," in *Proceedings of the 2nd conference on Symposium on Networked Systems Design & Implementation-Volume 2*, 2005, pp. 273–286.

[42] N. Loutas, E. Kamateri, F. Bosi, and K. Tarabanis, "Cloud computing interoperability: the state of play," in *2011 IEEE Third International Conference on Cloud Computing Technology and Science.* IEEE, 2011, pp. 752–757.

[43] N. Santos, K. P. Gummadi, and R. Rodrigues, "Towards trusted cloud computing." *HotCloud*, vol. 9, no. 9, p. 3, 2009.

[44] J. Hamilton, "Cooperative expendable micro-slice servers (cems): low cost, low power servers for internet-scale services," in *Conference on Innovative Data Systems Research.* Citeseer, 2009.

[45] N. Vasić, M. Barisits, V. Salzgeber, and D. Kostic, "Making cluster applications energy-aware," in *Proceedings of the 1st workshop on Automated control for Datacenters and Clouds*, 2009, pp. 37–42.

[46] S.-K. Chong, J. Abawajy, M. Ahmad, and I. R. A. Hamid, "Enhancing trust management in cloud environment," *Procedia-Social and Behavioral Sciences*, vol. 129, pp. 314–321, 2014.

[47] H. Baron, S. Heide, S. Mahmud, J. Yeoh, "Cloud security complexity: Challenges in managing security in hybrid and multi-cloud environments," Cloud Security Alliance, Tech. Rep., 2019.

[48] M. Mimiso, *Virtual Machine Escape Exploit Targets Xen*, Sept 2012. [Online]. Available: http://threatpost.com/virtual-machine-escape-exploit-targets-xen-090612/76979

[49] M. Dekker, D. Liveri, and M. Lakka, “Cloud Security Incident Reporting, Framework for Reporting about Major Cloud Security Incidents,” ENISA, Tech. Rep. doi:10.2788/14231, Dec 2013.

[50] J. Dee, *Amazon cloud infested with DDoS botnets*, July 2014. [Online]. Available: http://www.techwalls.com/amazon-cloud-infested-ddos-botnets/

[51] www.cyber edge.com, “Cyber threat defense report,” CYBEREDGE Group, Tech. Rep., 2014.

[52] J. Bourne, *Code Spaces RIP: Code hosting provider ceases trading after well orchestrated DDoS attack*, May 2014. [Online]. Available: http://www.cloudcomputing-news.net/news/2014/jun/19/code-spaces-rip-code-hosting-provider-ceases-trading-after-well-orchestrated-ddos-attack/

[53] B. Nahorney, “Symantec Intelligence Report,” Symantec, Tech. Rep., June 2015.

[54] P. Wood, “Internet Security Threat Report,” Symantec, Tech. Rep., April 2016.

[55] cisco.com, “Cisco Annual Cyber Security Report,” CISCO, Tech. Rep., 2017.

[56] C. Nenzani, “Csa annual report, 2018,” CSA, Tech. Rep., 2018. [Online]. Available: https://www.csagroup.org/annual-report/

[57] KrebsonSecurity, *QuickBooks Cloud Hosting Firm iNSYNQ Hit in Ransomware Attack*, 2019. [Online]. Available: https://krebsonsecurity.com/2019/07/quickbooks-cloud-hosting-firm-insynq-hit-in-ransomware-attack/

[58] Y. Zhang, A. Juels, M. K. Reiter, and T. Ristenpart, “Cross-vm side channels and their use to extract private keys,” in *ACM conference on Computer and communications security*. ACM, 2012, pp. 305–316.

[59] S. Gupta, and P. Kumar, “Taxonomy of cloud security,” *International Journal of Computer Science, Engineering and Applications*, vol. 3, no. 5, 2013.

[60] Z. Xiao, and Y. Xiao, “Security and privacy in cloud computing,” *IEEE Communications Surveys & Tutorials*, vol. 15, no. 2, pp. 843–859, 2013.

[61] C. Hoff, *Cloud computing security: From DDoS (distributed denial of service) to EDoS (economic denial of sustainability)*, 2008. [Online]. Available: http://www.rationalsurvivability.com/blog/?p=66.

[62] T. Jager, and J. Somorovsky, “How to break xml encryption,” in *Proceedings of the 18th ACM Conference on Computer and Communications Security*. ACM, 2011, pp. 413–422.

[63] AWS, *Aws management console*, 2019. [Online]. Available: http://aws.amazon.com/console/

[64] C. Modi, D. Patel, B. Borisaniya, A. Patel, and M. Rajarajan, “A survey on security issues and solutions at different layers of cloud computing,” *The Journal of Supercomputing*, vol. 63, no. 2, pp. 561–592, 2013.

[65] Y. Zhang, A. Juels, M. K. Reiter, and T. Ristenpart, "Cross-tenant side-channel attacks in paas clouds," in *Proceedings of the 2014 ACM SIGSAC Conference on Computer and Communications Security.* ACM, 2014, pp. 990–1003.

[66] S. A. Hussain, M. Fatima, A. Saeed, I. Raza, and R. K. Shahzad, "Multilevel classification of security concerns in cloud computing," *Applied Computing and Informatics*, vol. 13, no. 1, pp. 57–65, 2017.

[67] D. A. Fernandes, L. F. Soares, J. V. Gomes, M. M. Freire, and P. R. Inácio, "Security issues in cloud environments: a survey," *International Journal of Information Security*, vol. 13, no. 2, pp. 113–170, 2014.

[68] H. Ma, H. Ding, Y. Yang, Z. Mi, J. Y. Yang, and Z. Xiong, "Bayes-based arp attack detection algorithm for cloud centers," *Tsinghua Science and Technology*, vol. 21, no. 1, pp. 17–28, 2016.

[69] T. Hwang, Y. Shin, K. Son, and H. Park, "Design of a hypervisor-based rootkit detection method for virtualized systems in cloud computing environments," in *International Conference on Engineering and Technology.* Atlantis Press, 2013.

[70] U. Tupakula, V. Varadharajan, and N. Akku, "Intrusion detection techniques for infrastructure as a service cloud," in *IEEE Ninth International Conference on Dependable, Autonomic and Secure Computing (DASC).* IEEE, 2011, pp. 744–751.

[71] G. Wang, Z. Estrada, C. Pham, Z. Kalbarczyk, and R. Iyer, "Hypervisor introspection: Exploiting timing side-channels against vm monitoring," in *44th Annual IEEE/IFIP International Conference on Dependable Systems and Networks*, 2014, pp. 23–26.

[72] Citrix, *Download - XenServer*, May 2016. [Online]. Available: http://xenserver.org/open-source-virtualization-download.html

[73] VMware, *VMware vSphere*, 2017. [Online]. Available: https://my.vmware.com/en/web/vmware/info/slug/datacenter_cloud_infrastructure/vmware_vsphere/6_5

[74] vmware, *Download VMware Workstation Pro 12.5*, 2017. [Online]. Available: http://www.vmware.com/products/workstation/workstation-evaluation.html

[75] Orcle, *Download Oracle VirtualBox*, 2017. [Online]. Available: https://www.virtualbox.org/wiki/Downloads

[76] D. Milojičić, I. M. Llorente, and R. S. Montero, "Opennebula: A cloud management tool," *IEEE Internet Computing*, vol. 15, no. 2, pp. 11–14, 2011.

[77] H. P. Enterprise, *Download HPE Helion Eucalyptus*, April 2016. [Online]. Available: http://www8.hp.com/in/en/cloud/helion-eucalyptus-downloads.html#C4

[78] M. M. S. M. Habib, V. Varadharajan, "A framework for evaluating trust of service providers in cloud marketplaces," in *28th ACM Symp. Applied Computing, Coimbra, Portugal*, 2013, pp. 1–3.

[79] D. Slamanig, "Dynamic accumulator based discretionary access control for outsourced storage with unlinkable access," in *International Conference on Financial Cryptography and Data Security.* Springer, 2012, pp. 215–222.

[80] M. Raykova, H. Zhao, and S. M. Bellovin, "Privacy enhanced access control for outsourced data sharing," in *International Conference on Financial Cryptography and Data Security*. Springer, 2012, pp. 223–238.

[81] A. Yasinsac, and C. Irvine, "Help! is there a trustworthy-systems doctor in the house?" *IEEE Security Privacy*, vol. 11, no. 1, pp. 73–77, 2013.

[82] F. Curbera, M. Duftler, R. Khalaf, W. Nagy, N. Mukhi, and S. Weerawarana, "Unraveling the web services web: an introduction to soap, wsdl, and uddi," *IEEE Internet computing*, vol. 6, no. 2, pp. 86–93, 2002.

[83] F. Callegati, W. Cerroni, and M. Ramilli, "Man-in-the-middle attack to the https protocol," *IEEE Security & Privacy*, vol. 7, no. 1, pp. 78–81, 2009.

[84] Simplilearn, *ITIL: Key Concepts and Summary*, Mart 2019. [Online]. Available: https://www.simplilearn.com/itil-key-concepts-and-summary-article

[85] E. W. Bernroider, and M. Ivanov, "It project management control and the control objectives for it and related technology (cobit) framework," *International Journal of Project Management*, vol. 29, no. 3, pp. 325–336, 2011.

[86] M. Brenner, T. Schaaf, and A. Scherer, "Towards an information model for itil and iso/iec 20000 processes," in *2009 IFIP/IEEE International Symposium on Integrated Network Management*. IEEE, 2009, pp. 113–116.

[87] K. S. Brackney, and G. L. Helms, "A survey of attestation practices," *Auditing*, vol. 15, no. 2, p. 85, 1996.

[88] S. Cordero, *Cloud Controls Matrix Working Group*, Dec 2018. [Online]. Available: https://cloudsecurityalliance.org/working-groups/cloud-controls-matrix/#_overview

[89] G. Brunette and R. Mogull, "Security guidance for critical areas of focus in cloud computing v2.1," Cloud Security Alliance, Tech. Rep., 2009.

[90] N. C. C. S. W. Group *et al.*, "Nist cloud computing security reference architecture," Tech. Rep., 2013.

[91] B. Grobauer, T. Walloschek, and E. Stocker, "Understanding cloud computing vulnerabilities," *IEEE Security & Privacy*, vol. 9, no. 2, pp. 50–57, 2011.

[92] A. Ghorbel, M. Ghorbel, and M. Jmaiel, "Privacy in cloud computing environments: a survey and research challenges," *The Journal of Supercomputing*, vol. 73, no. 6, pp. 2763–2800, 2017.

[93] B. Deebak and F. Al-Turjman, "Robust lightweight privacy-preserving and session scheme interrogation for fog computing systems," *Journal of Information Security and Applications*, vol. 58, p. 102689, 2021.

[94] S. Belguith, N. Kaaniche, M. Laurent, A. Jemai, and R. Attia, "Accountable privacy preserving attribute based framework for authenticated encrypted access in clouds," *Journal of Parallel and Distributed Computing*, vol. 135, pp. 1–20, 2020.

[95] H. Liu, "A new form of dos attack in a cloud and its avoidance mechanism," in *Proceedings of the 2010 ACM Workshop on Cloud Computing Security Workshop*, 2010, pp. 65–76.

[96] R. Chow, P. Golle, M. Jakobsson, E. Shi, J. Staddon, R. Masuoka, and J. Molina, "Controlling data in the cloud: outsourcing computation without outsourcing control," in *Proceedings of the 2009 ACM Workshop on Cloud Computing Security*, 2009, pp. 85–90.

[97] M. Armbrust, A. Fox, R. Griffith, A. D. Joseph, R. Katz, A. Konwinski, G. Lee, D. Patterson, A. Rabkin, I. Stoica *et al.*, "A view of cloud computing," *Communications of the ACM*, vol. 53, no. 4, pp. 50–58, 2010.

[98] S. VivinSandar, and S. Shenai, "Economic denial of sustainability (EDoS) in cloud services using available: HTTP and XML based ddos attacks," *International Journal of Computer Applications*, vol. 41, no. 20, 2012.

[99] S. R. Krishna, and B. P. Rani, "Virtualization Security Issues and Mitigations in Cloud Computing," in *2nd Int. Conf. on Computational Intelligence and Informatics, Hyderabad, India*, 2017, pp. 117–128.

[100] C. Metz, *Amazon outage spans clouds 'insulated' from each other*, Apr 2011, Available: http://www.theregister.co.uk/2011/04/21/amazon_web_services_outages_spans_zones/.

[101] S. Subashini, and V. Kavitha, "A survey on security issues in service delivery models of cloud computing," *Journal of Network and Computer Applications*, vol. 34, no. 1, pp. 1–11, 2011.

[102] J. Li, X. Chen, M. Li, J. Li, P. P. Lee, and W. Lou, "Secure deduplication with efficient and reliable convergent key management," *IEEE Trans. on Parallel and Distributed Systems*, vol. 25, no. 6, pp. 1615–1625, 2014.

[103] I. Indu, P. R. Anand, and V. Bhaskar, "Identity and access management in cloud environment: Mechanisms and challenges," *Engineering Science and Technology, an International Journal*, vol. 21, no. 4, pp. 574–588, 2018.

[104] E. Maler, P. Mishra, and R. Philpott, "Assertions and Protocol for the Oasis Security Assertion Markup Language (SAML)," *OASIS*, 2003.

[105] E. E. Mon, and T. T. Naing, "The privacy-aware access control system using attribute-and role-based access control in private cloud," in *4th IEEE Int. Conf. on Broadband Network and Multimedia Technology (IC-BNMT), Beijing, China*, 2011, pp. 447–451.

[106] CSA, "The treacherous 12: Cloud computing top threats in 2016," Cloud Security Alliance, Tech. Rep. 500-291, Feb 2016.

[107] ISO, *ISO 38500 IT Governance Standard*, May 2008. [Online]. Available: http://www.38500.org/

[108] ISACA, *COBIT 5: Regulatory and Compliance*, april 2012. [Online]. Available: https://cobitonline.isaca.org/

[109] S. Cordero, *Cloud Controls Matrix Working Group*, June 2016. [Online]. Available: https://cloudsecurityalliance.org/group/cloud-controls-matrix/

[110] HHS.gov, *Health Information Privacy: Enforcement Highlights*, August 1996. [Online]. Available: https://www.hhs.gov/hipaa/index.html

[111] L. PCI Security Standards Council, *PCI Security*, 2006. [Online]. Available: https://www.pcisecuritystandards.org/pci_security/

[112] NIST, *Federal Information Security Modernization Act (FISMA) Implementation*, 2014. [Online]. Available: http://csrc.nist.gov/groups/SMA/fisma/index.html

[113] soxlaw.com, *The Sarbanes-Oxley Act*, 2002. [Online]. Available: https://www.ftc.gov/tips-advice/business-center/privacy-and-security/gramm-leach-bliley-act

[114] K. M. Khan, and Q. Malluhi, "Establishing trust in cloud computing," *IT Professional*, vol. 12, no. 5, pp. 20–27, 2010.

[115] M. K. Srinivasan, K. Sarukesi, P. Rodrigues, M. S. Manoj, and P. Revathy, "State-of-the-art cloud computing security taxonomies: a classification of security challenges in the present cloud computing environment," in *Int. Conf. on Advances in Computing, Communications and Informatics, Chennai, India*, 2012, pp. 470–476.

[116] F. Zhang, J. Wang, K. Sun, and A. Stavrou, "Hypercheck: A hardware-assisted integrity monitor," *IEEE Transactions on Dependable and Secure Computing*, vol. 11, no. 4, pp. 332–344, 2014.

[117] M. Pearce, S. Zeadally, and R. Hunt, "Virtualization: Issues, security threats, and solutions," *ACM Computing Surveys (CSUR)*, vol. 45, no. 2, pp. 1–39, 2013.

[118] A. Aviram, S. Hu, B. Ford, and R. Gummadi, "Determinating timing channels in compute clouds," in *Proceedings of the 2010 ACM Workshop on Cloud Computing Security Workshop*, 2010, pp. 103–108.

[119] M. Jensen, J. Schwenk, N. Gruschka, and L. L. Iacono, "On technical security issues in cloud computing," in *2009 IEEE International Conference on Cloud Computing*. IEEE, 2009, pp. 109–116.

[120] M. Almorsy, J. Grundy, and I. Müller, "An analysis of the cloud computing security problem," *arXiv preprint arXiv:1609.01107*, 2016.

[121] S. Pearson, "Privacy, security and trust in cloud computing," in *Privacy and Security for Cloud Computing*. Springer, 2013, pp. 3–42.

[122] R. Charanya, M. Aramudhan, K. Mohan, and S. Nithya, "Levels of security issues in cloud computing," *International Journal of Engineering and Technology*, vol. 5, no. 2, pp. 1912–1920, 2013.

[123] S. Sengupta, V. Kaulgud, and V. S. Sharma, "Cloud computing security–trends and research directions," in *2011 IEEE World Congress on Services*. IEEE, 2011, pp. 524–531.

[124] S. Yu, C. Wang, K. Ren, and W. Lou, "Achieving secure, scalable, and fine-grained data access control in cloud computing," in *IEEE INFOCOM*. IEEE, 2010, pp. 1–9.

[125] C. Park, H. Kim, D. Hong, and C. Seo, "A symmetric key based deduplicatable proof of storage for encrypted data in cloud storage environments," *Security and Communication Networks*, vol. 2018, pp. 1–12, 2018.

[126] Y. K. Sinjilawi, M. Q. Al-Nabhan, and E. A. Abu-Shanab, "Addressing security and privacy issues in cloud computing." *Journal of Emerging Technologies in Web Intelligence*, vol. 6, no. 2, 2014.

[127] B. R. Kandukuri, A. Rakshit *et al.*, "Cloud security issues," in *2009 IEEE International Conference on Services Computing*. IEEE, 2009, pp. 517–520.

[128] B. R. Rowe *et al.*, "Will outsourcing it security lead to a higher social level of security?" pp. 1–6, 2008.

[129] M. D. Dikaiakos, D. Katsaros, P. Mehra, G. Pallis, and A. Vakali, "Cloud computing: Distributed internet computing for it and scientific research," *IEEE Internet Computing*, vol. 13, no. 5, pp. 10–13, 2009.

[130] K. Fatema, P. D. Healy, V. C. Emeakaroha, J. P. Morrison, and T. Lynn, "A user data location control model for cloud services." in *CLOSER*, 2014, pp. 476–488.

[131] T. J. Neela and N. Saravanan, "Privacy preserving approaches in cloud: a survey," *Indian Journal of Science and Technology*, vol. 6, no. 5, pp. 4531–4535, 2013.

[132] CloudVPS, *The open cloud*, 2011. [Online]. Available: https://www.cloudvps.com/openstack

[133] J.-M. Kim, H.-Y. Jeong, I. Cho, S. M. Kang, and J. H. Park, "A secure smart-work service model based OpenStack for cloud computing," *Cluster Computing*, vol. 17, no. 3, pp. 691–702, 2014.

[134] VEXXHOST, *VEXXHOST: OpenStack Public Cloud*, 2015. [Online]. Available: https://vexxhost.com/public-cloud

[135] AURO, *AURO Public Cloud*, 2016. [Online]. Available: https://auro.io/public_cloud_hosting/overview

[136] Rackspace, *Introducing the Rackspace cloud*, 2015. [Online]. Available: https://www.rackspace.com/en-in/library/what-is-a-private-cloud

[137] M. Kelly, *DataCentred World First: OpenStack Public Cloud on 64-bit ARM Servers*, 2014. [Online]. Available: http://www.datacentred.co.uk/news/datacentred-world-first-openstack-public-cloud-on-64-bit-arm-servers/

[138] Elastx, *About Elastx*, 2012. [Online]. Available: https://elastx.se/en/about-elastx

[139] OpenStack, *OpenStack Marketplace: Dualtec Public Cloud*, 2011. [Online]. Available: https://www.openstack.org/marketplace/public-clouds/uol-diveo/dualtec-public-cloud

[140] Internap, *AgileCLOUD: High-performance cloud Infrastructure-as-a-Service powered by OpenStack*, 2016. [Online]. Available: http://www.internap.com/cloud/public-cloud/

[141] B. B. Group, *IBM Blue Box Continues its Path Toward Keeping Your Cloud Environment Safe*, Oct 2016. [Online]. Available: https://www.blueboxcloud.com/insight/blog-article/ibm-blue-box-continues-its-path-toward-keeping-your-cloud-environment-safe

[142] Platform9, *Platform9 Announces Hybrid Cloud-as-a-Service for OpenStack Community*, Oct 2016. [Online]. Available: https://platform9.com/press/openstack-hybrid/

[143] M. Jouini, L. B. A. Rabai, and A. B. Aissa, "Classification of security threats in information systems." *Procedia Computer Science*, vol. 32, pp. 489–496, 2014.

[144] P. Mishra, E. S. Pilli, V. Varadharajan, and U. Tupakula, "Intrusion detection techniques in cloud environment: A survey," *Journal of Network and Computer Applications*, vol. 77, pp. 18–47, 2017.

[145] CVE, *CVE Details: The Ultimate Security Vulnerabiltiy Data Source*, 2007. [Online]. Available: https://www.cvedetails.com/cve/CVE-2007-4593/

[146] A. Nguyen, *Raytheon hit by cloud-based phishing attack*, OCT 2011, http://www.techworld.com/news/security/raytheon-hit-by-cloud based-phishing-attack-3310603/.

[147] M. Pearce, S. Zeadally, and R. Hunt, "Virtualization: Issues, security threats, and solutions," *ACM Computing Surveys (CSUR)*, vol. 45, no. 2, p. 17, 2013.

[148] R. Wojtczuk, "Subverting the xen hypervisor," *Black Hat USA*, vol. 2008, 2008.

[149] P. Stewin, and I. Bystrov, "Understanding dma malware," in *Detection of Intrusions and Malware, and Vulnerability Assessment.* Springer, 2012, pp. 21–41.

[150] C. Kallenberg, J. Butterworth, X. Kovah, and C. Cornwell, "Defeating signed bios enforcement," *EkoParty, Buenos Aires*, 2013.

[151] P. Ferrie, "Attacks on more virtual machine emulators," *Symantec Technology Exchange*, vol. 55, 2007.

[152] M. Mulazzani, S. Schrittwieser, M. Leithner, M. Huber, and E. Weippl, "Dark clouds on the horizon: Using cloud storage as attack vector and online slack space." in *USENIX Security Symposium*, 2011, pp. 1–11.

[153] J. Oberheide, E. Cooke, and F. Jahanian, "Empirical exploitation of live virtual machine migration," in *Proc. of BlackHat DC Convention.* Citeseer, 2008.

[154] S. Sanfilippo, *HPING 3*, 2005. [Online]. Available: http://www.hping.org/hping3.html

[155] G. F. Lyon, *Nmap network scanning: The official Nmap project guide to network discovery and security scanning.* Insecure, 2009.

[156] V. Varadharajan and U. Tupakula, "Security as a service model for cloud environment," *IEEE Transactions on Network and Service Management*, vol. 11, no. 1, pp. 60–75, March 2014.

[157] KDD, *KDD Dataset*, 1998, http://kdd.ics.uci.edu/databases/kddcup99/ kddcup99.html.

[158] N. Moustafa and J. Slay, "UNSW-NB15: A Comprehensive Data Set for Network Intrusion Detection Systems (UNSW-NB15 Network Data Set)," in *Military Communications and Information Systems Conference (MilCIS), Canberra, Australia*, 2015, pp. 1–6.

[159] P. Mishra, E. S. Pilli, and R. Joshi, "Forensic analysis of e-mail date and time spoofing," in *2012 Third International Conference on Computer and Communication Technology.* IEEE, 2012, pp. 309–314.

[160] P. Mishra, E. S. Pilli, V. Varadharajan, and U. Tupakula, "Vaed: Vmi-assisted evasion detection approach for infrastructure as a service cloud," *Concurrency and Computation: Practice and Experience*, vol. 29, no. 12, p. e4133, 2017.

[161] P. Mishra, E. Pilli, V. Varadharajan, and U. Tupakula, "PSI-NetVisor: Program semantic aware intrusion detection at network and hypervisor layer in cloud", *Journal of Intelligent & Fuzzy Systems*, vol. 32, no. 4, pp. 2909–2921, 2017.

[162] S. S. Alarifi, and S. D. Wolthusen, "Detecting anomalies in iaas environments through virtual machine host system call analysis," in *International Conference on Internet Technology And Secured Transactions*. IEEE, 2012, pp. 211–218.

[163] M. I. Sharif, W. Lee, W. Cui, and A. Lanzi, "Secure in-vm monitoring using hardware virtualization," in *16th ACM conf. on Computer and Communications Security, Chicago, Illinois, USA*, 2009, pp. 477–487.

[164] H. Xiong, Z. Liu, W. Xu, and S. Jiao, "Libvmi: a library for bridging the semantic gap between guest os and vmm," in *IEEE 12th International Conference on Computer and Information Technology*. IEEE, 2012, pp. 549–556.

[165] B. D. Payne, M. De Carbone, and W. Lee, "Secure and flexible monitoring of virtual machines," in *Computer Security Applications Conference, Florida, USA*, 2007, pp. 385–397.

[166] T. K. Lengyel, S. Maresca, B. D. Payne, G. D. Webster, S. Vogl, and A. Kiayias, "Scalability, fidelity and stealth in the drakvuf dynamic malware analysis system," in *30th Annual Computer Security Applications Conference*. ACM, 2014, pp. 386–395.

[167] X. Jiang, X. Wang, and D. Xu, "Stealthy malware detection through VMM-based out-of-the-box semantic view reconstruction," in *14th ACM conf. on Computer and Communications Security, New York, NY, USA*, 2007, pp. 128–138.

[168] P. Mishra, V. Varadharajan, E. Pilli, and U. Tupakula, "Vmguard: A vmi-based security architecture for intrusion detection in cloud environment," *IEEE Transactions on Cloud Computing*, vol. Early access, pp. 1–14, 2018.

[169] P. Mishra, E. S. Pilli, V. Varadharajan, and U. Tupakula, "Vaed: Vmi-assisted evasion detection approach for infrastructure as a service cloud," *Concurrency and Computation: Practice and Experience*, vol. 29, no. 12, pp. 1–21, 2017.

[170] I. Gul, and M. Hussain, "Distributed cloud intrusion detection model," *International Journal of Advanced Science and Technology*, vol. 34, no. 38, p. 135, 2011.

[171] C. Modi and D. Patel, "A novel hybrid-network intrusion detection system (h-nids) in cloud computing," in *IEEE Symposium on Computational Intelligence in Cyber Security (CICS)*, April 2013, pp. 23–30.

[172] S. Gupta, and P. Kumar, "System cum program-wide lightweight malicious program execution detection scheme for cloud," *Information Security Journal: A Global Perspective*, vol. 23, no. 3, pp. 86–99, 2014.

[173] S. Bharadwaja, W. Sun, M. Niamat, and F. Shen, "Collabra: A Xen hypervisor based collaborative intrusion detection system," in *8th Int. Conf. on Information Technology: New Generations (ITNG), Las Vegas, Nevada, USA*, 2011, pp. 695–700.

[174] C. C. Lo, C. C. Huang, and J. Ku, "A cooperative intrusion detection system framework for cloud computing networks," in *39th Int. Conf. on Parallel Processing Workshops (ICPPW), San Diego, CA*, 2010, pp. 280–284.

[175] C. Maiero, and M. Miculan, "Unobservable intrusion detection based on call traces in paravirtualized systems," in *Int. Conf. on Security and Cryptography, Seville, Spain*, 2011, pp. 300–306.

[176] C. Ko, G. Fink, and K. Levitt, "Automated detection of vulnerabilities in privileged programs by execution monitoring," in *Computer Security Applications Conference, New Orleans, Louisiana*, 1994, pp. 134–144.

[177] S. Forrest, S. Hofmeyr, A. Somayaji, T. Longstaff *et al.*, "A sense of self for unix processes," in *IEEE Symposium on Security and Privacy, Oakland, CA, USA*, 1996, pp. 120–128.

[178] S. A. Hofmeyr, S. Forrest, and A. Somayaji, "Intrusion detection using sequences of system calls," *Journal of Computer Security*, vol. 6, no. 3, pp. 151–180, 1998.

[179] C. Warrender, S. Forrest, and B. Pearlmutter, "Detecting Intrusions using System Calls: Alternative Data Models," in *IEEE Symposium on Security and Privacy, Oakland, California*, 1999, pp. 133–145.

[180] D.-K. Kang, D. Fuller, and V. Honavar, "Learning classifiers for misuse and anomaly detection using a bag of system calls representation," in *IEEE Information Assurance Workshop, New York*, USA, 2005, pp. 118–125.

[181] Q.-B. Yin, L.-R. Shen, R.-B. Zhang, X.-Y. Li, and H.-Q. Wang, "Intrusion detection based on hidden Markov model," in *Int. Conf. on Machine Learning and Cybernetics, Xi'an, China*, 2003, pp. 3115–3118.

[182] S. Alarifi, and S. Wolthusen, "Anomaly detection for ephemeral cloud iaas virtual machines," in *Network and System Security*. Springer, 2013, pp. 321–335.

[183] P. Mishra, K. Khurana, S. Gupta, and M. K. Sharma, "Vmanalyzer: Malware semantic analysis using integrated cnn and bi-directional lstm for detecting vm-level attacks in cloud," in *2019 Twelfth International Conference on Contemporary Computing (IC3)*, 2019, pp. 1–6.

[184] T. Garfinkel, M. Rosenblum *et al.*, "A Virtual Machine Introspection Based Architecture for Intrusion Detection," in *Network and Distributed System Security Symposium, San Diego, California*, 2003, pp. 191–206.

[185] J. Arshad, P. Townend, and J. Xu, "A novel intrusion severity analysis approach for clouds," *Future Generation Computer Systems*, vol. 29, no. 1, pp. 416–428, 2013.

[186] M. Bernaschi, E. Gabrielli, and L. V. Mancini, "Remus: a security-enhanced operating system," *ACM Trans. on Information and System Security (TISSEC)*, vol. 5, no. 1, pp. 36–61, 2002.

[187] S. Roschke, F. Cheng, and C. Meinel, "Intrusion detection in the cloud," in *IEEE 8th International Conference on Dependable, Autonomic and Secure Computing*. IEEE, 2009, pp. 729–734.

[188] C. Mazzariello, R. Bifulco, and R. Canonico, "Integrating a network IDS into an open source cloud computing environment," in *6th Int. Conf. on Information Assurance and Security (IAS), Atlanta, GA, USA*, 2010, pp. 265–270.

[189] J.-H. Lee, M.-W. Park, J.-H. Eom, and T.-M. Chung, "Multi-level intrusion detection system and log management in cloud computing," in *13th Int. Conf. on Advanced Communication Technology (ICACT), Gangwon-Do, South Korea*, 2011, pp. 552–555.

[190] M. K. Srinivasan, K. Sarukesi, A. Keshava, and P. Revathy, "Ids tier-1 ux-engine subsystem design and implementation using self-organizing map (som) for secure cloud computing environment," in *Recent Trends in Computer Networks and Distributed Systems Security*. Springer, 2012, pp. 432–443.

[191] C. Modi, D. Patel, B. Borisanya, A. Patel, and M. Rajarajan, "A novel framework for intrusion detection in cloud," in *Fifth International Conference on Security of Information and Networks*. ACM, 2012, pp. 67–74.

[192] U. Tupakula, V. Varadharajan, and N. Akku, "Intrusion detection techniques for infrastructure as a service cloud," in *IEEE Ninth International Conference on Dependable, Autonomic and Secure Computing (DASC)*. IEEE, 2011, pp. 744–751.

[193] N. Pandeeswari, and G. Kumar, "Anomaly detection system in cloud environment using fuzzy clustering based ann," *Mobile Networks and Applications*, pp. 1–12, 2015.

[194] X. Ma, X. Fu, B. Luo, X. Du, and M. Guizani, "A design of firewall based on feedback of intrusion detection system in cloud environment," in *2019 IEEE Global Communications Conference (GLOBECOM)*. IEEE, 2019, pp. 1–6.

[195] C. Xu, R. Zhang, M. Xie, and L. Yang, "Network intrusion detection system as a service in openstack cloud," in *2020 International Conference on Computing, Networking and Communications (ICNC)*. IEEE, 2020, pp. 450–455.

[196] C.-M. Chen, D. Guan, Y.-Z. Huang, and Y.-H. Ou, "State-based attack detection for cloud," in *IEEE Int. Symposium on Next-Generation Electronics (ISNE), Kaohsiung, Taiwan*, 2013, pp. 177–180.

[197] S. Gupta and P. Kumar, "An immediate system call sequence based approach for detecting malicious program executions in cloud environment," *Wireless Personal Communications*, vol. 81, no. 1, pp. 405–425, 2015.

[198] V. Varadharajan, and U. Tupakula, "On the design and implementation of an integrated security architecture for cloud with improved resilience," *IEEE Trans. on Cloud Computing*, 2016.

[199] H. A. Kholidy, A. Erradi, S. Abdelwahed, and F. Baiardi, "Ha-cids: A hierarchical and autonomous ids for cloud systems," in *Fifth International Conference on Computational Intelligence, Communication Systems and Networks*. IEEE, 2013, pp. 179–184.

[200] Z. Al Haddad, M. Hanoune, and A. Mamouni, "A Collaborative Network Intrusion Detection System (C-NIDS) in Cloud Computing," *International Journal of Communication Networks and Information Security*, vol. 8, no. 3, p. 130, 2016.

[201] A. S. Abed, T. C. Clancy, and D. S. Levy, "Applying Bag of System Calls for Anomalous Behavior Detection of Applications in Linux Containers," in *IEEE Globecom Workshops (GC Wkshps), San Diego, CA, USA*, Dec. 2015, pp. 1–5.

[202] C. Benninger, S. W. Neville, Y. O. Yazır, C. Matthews, and Y. Coady, "Maitland: Lighter-weight vm introspection to support cyber-security in the cloud," in *IEEE 5th International Conference on Cloud Computing (CLOUD)*. IEEE, 2012, pp. 471–478.

[203] D. Singh, D. Patel, B. Borisaniya, and C. Modi, "Collaborative IDS Framework for Cloud," *International Journal of Network Security*, vol. 18, no. 4, pp. 699–709, 2016.

[204] M. R. Watson, A. K. Marnerides, A. Mauthe, D. Hutchison *et al.*, "Malware detection in cloud computing infrastructures," *IEEE Trans. on Dependable and Secure Computing*, vol. 13, no. 2, pp. 192–205, 2016.

[205] C.-W. Tien, H.-K. Pao, and C.-H. Lin, "Efficient and effective nids for cloud virtualization environment," in *IEEE 4th International Conference on Cloud Computing Technology and Science (CloudCom)*, 2012, pp. 249–254.

[206] S. Gupta, P. Kumar, and A. Abraham, "A profile based network intrusion detection and prevention system for securing cloud environment," *Int. J. of Distributed Sensor Networks*, vol. 2013, no. 364575, pp. 1–12, 2013.

[207] Y. Fu, and Z. Lin, "Space traveling across VM: Automatically bridging the semantic gap in virtual machine introspection via online kernel data redirection," in *IEEE Symposium on Security and Privacy, San Francisco, CA*, 2012, pp. 586–600.

[208] R. Wu, P. Chen, P. Liu, and B. Mao, "System Call Redirection: A Practical Approach to Meeting Real-World Virtual Machine Introspection Needs," in *44th Annual IEEE/IFIP Int. Conf. on Dependable Systems and Networks, Atlanta, GA, USA*, 2014, pp. 574–585.

[209] Z. Li, W. Sun, and L. Wang, "A neural network based distributed intrusion detection system on cloud platform," in *Cloud Computing and Intelligent Systems (CCIS), 2012 IEEE 2nd International Conference on*, vol. 1. IEEE, 2012, pp. 75–79.

[210] M. Ficco, L. Tasquier, and R. Aversa, "Intrusion detection in cloud computing," in *8th Int. Conf. on P2P, Parallel, Grid, Cloud and Internet Computing,Compiegne, France*, 2013, pp. 276–283.

[211] P. Garcia-Teodoro, J. Diaz-Verdejo, G. Maciá-Fernández, and E. Vázquez, "Anomaly-based network intrusion detection: Techniques, systems and challenges," *Computers & Security*, vol. 28, no. 1, pp. 18–28, 2009.

[212] Y. Hebbal, S. Laniepce, and J.-M. Menaud, "Virtual Machine Introspection: Techniques and Applications," in *10th Int. Conf. on Availability, Reliability and Security (ARES), Toulouse, France*, 2015, pp. 676–685.

[213] D. Das, and S. Gupta, "Towards a novel cross-media encryption-cum-obfuscation technique," in *Information, Photonics and Communication*. Springer, 2020, pp. 77–85.

[214] P. Mishra, I. Verma, S. Gupta, V. S. Rana, and K. Kadarla, "vproval: Introspection based process validation for detecting malware in kvm-based cloud environment," in *2019 Fourth International Conference on Fog and Mobile Edge Computing (FMEC)*, 2019, pp. 271–277.

[215] F. Zhang, J. Chen, H. Chen, and B. Zang, "Cloudvisor: retrofitting protection of virtual machines in multi-tenant cloud with nested virtualization," in *Proceedings of the Twenty-Third ACM Symposium on Operating Systems Principles*, 2011, pp. 203–216.

[216] Y. Xia, Y. Liu, and H. Chen, "Architecture support for guest-transparent vm protection from untrusted hypervisor and physical attacks," in *2013 IEEE 19th International Symposium on High Performance Computer Architecture (HPCA)*. IEEE, 2013, pp. 246–257.

[217] V. Varadharajan and U. Tupakula, "Security as a service model for cloud environment," *IEEE Transactions on Network and Service Management*, vol. 11, no. 1, pp. 60–75, March 2014.

[218] A.-S. K. Pathan, *The state of the art in intrusion prevention and detection*. CRC Press, 2014.

[219] C. V. Kopek, E. W. Fulp, and P. S. Wheeler, "Distributed data parallel techniques for content-matching intrusion detection systems," in *Military Communications Conference*. IEEE, 2007, pp. 1–7.

[220] L. Vokorokos, M. Ennert, M. Cajkovsky, and A. Turínska, "A distributed network intrusion detection system architecture based on computer stations using gpgpu," in *17th International Conference on Intelligent Engineering Systems (INES)*. IEEE, 2013, pp. 323–326.

[221] G. Kim, S. Lee, and S. Kim, "A novel hybrid intrusion detection method integrating anomaly detection with misuse detection," *Expert Systems with Applications*, vol. 41, no. 4, pp. 1690–1700, 2014.

[222] F. Amiri, M. R. Yousefi, C. Lucas, A. Shakery, and N. Yazdani, "Mutual information-based feature selection for intrusion detection systems," *Journal of Network and Computer Applications*, vol. 34, no. 4, pp. 1184–1199, 2011.

[223] A. Tajbakhsh, M. Rahmati, and A. Mirzaei, "Intrusion detection using fuzzy association rules," *Applied Soft Computing*, vol. 9, no. 2, pp. 462–469, 2009.

[224] Wenzel, *Nitro - github*, 2017. [Online]. Available: https://github.com/KVM-VMI/nitro

[225] Hex-Rays, IDA Pro 7.6, 2021. [Online]. Available: https://hex-rays.com/.

[226] P. Garcia-Teodoro, J. Diaz-Verdejo, G. Maciá-Fernández, and E. Vázquez, "Anomaly-based network intrusion detection: Techniques, systems and challenges," *computers & security*, vol. 28, no. 1, pp. 18–28, 2009.

[227] P. Kumar, N. Nitin, V. Sehgal, K. Shah, S. S. P. Shukla, and D. S. Chauhan, "A novel approach for security in cloud computing using hidden Markov model and clustering," in *World Congress on Information and Communication Technologies (WICT)*. IEEE, 2011, pp. 810–815.

[228] L. Khan, M. Awad, and B. Thuraisingham, "A new intrusion detection system using support vector machines and hierarchical clustering," *The VLDB Journal*, vol. 16, no. 4, pp. 507–521, 2007.

[229] S. Forrest, S. Hofmeyr, A. Somayaji, T. Longstaff *et al.*, "A sense of self for unix processes," in *IEEE Symposium on Security and Privacy*. IEEE, 1996, pp. 120–128.

[230] S. A. Hofmeyr, S. Forrest, and A. Somayaji, "Intrusion detection using sequences of system calls," *Journal of Computer Security*, vol. 6, no. 3, pp. 151–180, 1998.

[231] UNM. (1998) Unm dataset. [Online]. Available: http://www.cs.unm.edu/immsec/systemcalls.htm

[232] C. Warrender, S. Forrest, and B. Pearlmutter, "Detecting intrusions using system calls: Alternative data models," in *IEEE Symposium on Security and Privacy*. IEEE, 1999, pp. 133–145.

[233] D.-K. Kang, D. Fuller, and V. Honavar, "Learning classifiers for misuse and anomaly detection using a bag of system calls representation," in *Information Assurance Workshop, Sixth Annual IEEE SMC*. IEEE, 2005, pp. 118–125.

[234] S.-B. Cho, and H.-J. Park, "Efficient anomaly detection by modeling privilege flows using hidden Markov model," *Computers & Security*, vol. 22, no. 1, pp. 45–55, 2003.

[235] Q.-B. Yin, L.-R. Shen, R.-B. Zhang, X.-Y. Li, and H.-Q. Wang, "Intrusion detection based on hidden Markov model," in *International Conference on Machine Learning and Cybernetics*, vol. 5. IEEE, 2003, pp. 3115–3118.

[236] W. Lee, S. J. Stolfo, and P. K. Chan, "Learning patterns from unix process execution traces for intrusion detection," in *AAAI Workshop on AI Approaches to Fraud Detection and Risk Management*, 1997, pp. 50–56.

[237] C. Ko, G. Fink, and K. Levitt, "Automated detection of vulnerabilities in privileged programs by execution monitoring," in *Computer Security Applications Conference*. IEEE, 1994, pp. 134–144.

[238] D. Yuxin, Y. Xuebing, Z. Di, D. Li, and A. Zhanchao, "Feature representation and selection in malicious code detection methods based on static system calls," *Computers & Security*, vol. 30, no. 6, pp. 514–524, 2011.

[239] M. G. Schultz, E. Eskin, E. Zadok, and S. J. Stolfo, "Data mining methods for detection of new malicious executables," in *IEEE Symposium on Security and Privacy*. IEEE, 2001, pp. 38–49.

[240] M. M. Masud, L. Khan, and B. Thuraisingham, "A hybrid model to detect malicious executables," in *IEEE International Conference on Communications*. IEEE, 2007, pp. 1443–1448.

[241] S. Garfinkel, *Architects of the Information Society: 35 Years of the Laboratory for Computer Science at MIT*, 1st ed. Cambridge, Massachusetts: MIT Press, 1999.

[242] T. K. Lengyel *et al.*, "Scalability, fidelity and stealth in the drakvuf dynamic malware analysis system," in *30th Annual Computer Security Applications Conf., New York, NY, USA*, 2014, pp. 386–395.

[243] A. Dinaburg *et al.*, "Ether: Malware Analysis via Hardware Virtualization Extensions," in *15th ACM Conf. on Computer and communications security, VA, USA*, 2008, pp. 51–62.

[244] M. Ben-Yehuda, M. D. Day, Z. Dubitzky, M. Factor, N. Har'El, A. Gordon, A. Liguori, O. Wasserman, and B.-A. Yassour, "The turtles project: Design and implementation of nested virtualization." in *OSDI*, vol. 10, 2010, pp. 423–436.

[245] H. Moon, H. Lee, J. Lee, K. Kim, Y. Paek, and B. B. Kang, "Vigilare: toward snoop-based kernel integrity monitor," in *ACM conference on Computer and communications security*. ACM, 2012, pp. 28–37.

[246] E. Keller, J. Szefer, J. Rexford, and R. B. Lee, "Nohype: virtualized cloud infrastructure without the virtualization," in *ACM SIGARCH Computer Architecture News*, vol. 38, no. 3. ACM, 2010, pp. 350–361.

[247] B. Ding, Y. He, Y. Wu, and Y. Lin, "Hyperverify: a vm-assisted architecture for monitoring hypervisor non-control data," in *7th International Conference on Software Security and Reliability-Companion (SERE-C)*. IEEE, 2013, pp. 26–34.

[248] C. N. Modi, D. R. Patel, A. Patel, and M. Rajarajan, "Integrating signature apriori based network intrusion detection system (nids) in cloud computing," *Procedia Technology*, vol. 6, pp. 905–912, 2012.

[249] J. Shi, Y. Yang, and C. Tang, "Hardware assisted hypervisor introspection," *SpringerPlus*, vol. 5, no. 1, p. 647, 2016.

[250] D. Zissis and D. Lekkas, "Addressing cloud computing security issues," *Future Generation Computer Systems*, vol. 28, no. 3, pp. 583–592, 2012.

[251] H. Wu, Y. Ding, C. Winer, and L. Yao, "Network security for virtual machine in cloud computing," in *5th International Conference on Computer Sciences and Convergence Information Technology*. IEEE, 2010, pp. 18–21.

[252] INFOSEC, *Xoic Download website.*

[253] Daniel, *Rudy Download website*, 2016. [Online]. Available: https://github.com/DanielRTeixeira/R.U.D.Y.

[254] STORMSECURITY, *DDoSIM Download website*, 2010. [Online]. Available: https://stormsecurity.wordpress.com/2009/03/03/application-layer-ddos-simulator/

[255] S. T. King, and P. M. Chen, "Subvirt: Implementing malware with virtual machines," in *2006 IEEE Symposium on Security and Privacy (S&P'06)*. IEEE, 2006, pp. 14–pp.

[256] T. Wilson, *Injector github*, 2016. [Online]. Available: https://github.com/nettitude/DLLInjection/blob/master/Nettitude/Injector.h

[257] N. Admin, *nessus website*, 1998. [Online]. Available: https://www.tenable.com/products/nessus/nessus-professional

[258] R. Beede, *Netcat downlaod*, 2014. [Online]. Available: https://github.com/diegocr/netcat

[259] ossec.net, *OSSEC Download*, 2021. [Online]. Available: https://www.ossec.net/

[260] AWS, *Prowler github*, 2019. [Online]. Available: https://github.com/toniblyx/prowler

[261] Netflix, *Security Monkey github*, 2019. [Online]. Available: https://github.com/Netflix/security_monkey

[262] RiotGames, *cloud-inquisitor github*, 2019. [Online]. Available: https://github.com/RiotGames/cloud-inquisitor

[263] P. Technologies, *Loic Download*, 2009. [Online]. Available: https://github.com/NewEraCracker/LOIC

[264] Akubera, *hoic Download*, 2020. [Online]. Available: https://github.com/akubera/hoic

[265] B. Shteiman, *Hulk github*, 2014. [Online]. Available: https://github.com/grafov/hulk

[266] Soft112, *Pyloris Download website.*

[267] H. D. Moore, *Metasploit website*, 2021. [Online]. Available: https://www.metasploit.com/

[268] C. R. Russel, *dSniff Download website*, 2001. [Online]. Available: http://www.ouah.org/dsniffintr.htm

[269] M. Stampar, *Eternul Rocks github*, 2017. [Online]. Available: https://github.com/stamparm/EternalRocks

[270] E. Maland, *The Fat Rat github*, 2021. [Online]. Available: https://github.com/Screetsec/TheFatRat

[271] TSA, *Adware Virus Removal Tool*, 2019. [Online]. Available: https://www.adwareremovaltool.org/

[272] Malwarebytes, *Spyware*, 2019. [Online]. Available: https://www.malwarebytes.com/spyware/

[273] G. E. Hall, *Riskware*, 2019. [Online]. Available: https://www.2-spyware.com/remove-riskware-ifeohijack.html

[274] Kaspersky, *Hoax (Software)*, 2019. [Online]. Available: https://encyclopedia.kaspersky.com/knowledge/hoax/

[275] Milenkoski, *HInjector*, 2013. [Online]. Available: https://github.com/hinj/hInjector

[276] A. Milenkoski, B. Payne, N. Antunes, M. Vieira, and S. Kounev, "Hinjector: Injecting hypercall attacks for evaluating vmi-based intrusion detection systems," in *Annual Computer Security Applications Conference (ACSAC 2013)*, 2013, pp. 1–2.

[277] M. Montoro, *CAIN and ABEL*, 2014. [Online]. Available: https://web.archive.org/web/20190603235413/http://www.oxid.it/cain.html

[278] ScoutSuite@nccgroup.com, *Scout github*, 2019. [Online]. Available: https://github.com/nccgroup/ScoutSuite

[279] A. Wave, *Cloud Sploit github*, 2021. [Online]. Available: https://github.com/cloudsploit/scans

[280] K. Thangavelu, *Cloud Custodian github*, 2019. [Online]. Available: https://github.com/capitalone/cloud-custodian

[281] J. Naglieri, *Stream Alert github*, 2019. [Online]. Available: https://github.com/airbnb/streamalert

[282] D. Jones, *Hammer github*, 2019. [Online]. Available: https://github.com/dowjones/hammer

[283] Rekall-Forensic.com, *Rekall github*, 2020. [Online]. Available: https://github.com/google/rekall

[284] ——, *Volatility*, 2021. [Online]. Available: https://github.com/volatilityfoundation/volatility

[285] A. M. Azab, P. Ning, Z. Wang, X. Jiang, X. Zhang, and N. C. Skalsky, "Hypersentry: enabling stealthy in-context measurement of hypervisor integrity," in *Proceedings of the 17th ACM conference on Computer and communications security*, 2010, pp. 38–49.

[286] J. Shi, Y. Yang, C. Li, and X. Wang, "Spems: A stealthy and practical execution monitoring system based on vmi," in *International Conference on Cloud Computing and Security*. Springer, 2015, pp. 380–389.

[287] B. D. Payne, M. Carbone, M. Sharif, and W. Lee, "Lares: An architecture for secure active monitoring using virtualization," in *IEEE Symposium on Security and Privacy, California, USA*, 2008, pp. 233–247.

[288] K. Kourai and S. Chiba, "HyperSpector: virtual distributed monitoring environments for secure intrusion detection," in *1st ACM Int. Conf. on Virtual Execution Environments, Chicago, IL, USA*, 2005, pp. 197–207.

[289] F. Tang, B. Ma, J. Li, F. Zhang, J. Su, and J. Ma, "Ransomspector: An introspection-based approach to detect crypto ransomware," *Computers & Security*, vol. 97, p. 101997, 2020.

[290] P. Mishra, E. S. Pilli, V. Varadharajan, and U. Tupakula, "Out-vm monitoring for malicious network packet detection in cloud," in *ISEA Asia Security and Privacy (ISEASP)*. IEEE, 2017, pp. 1–10.

[291] D. Zhan, H. Li, L. Ye, H. Zhang, B. Fang, and X. Du, "A low-overhead kernel object monitoring approach for virtual machine introspection," in *IEEE International Conference on Communications (ICC)*. IEEE, 2019, pp. 1–6.

[292] P. Mishra, P. Aggarwal, A. Vidyarthi, P. Singh, B. Khan, H. Haes Alhelou, and P. Siano, "Vmshield: Memory introspection-based malware detection to secure cloud-based services against stealthy attacks," *IEEE Transactions on Industrial Informatics*, vol. Early Access, pp. 1–12, 2021.

[293] D. Kirat, G. Vigna, and C. Kruegel, "Barecloud: Bare-metal analysis-based evasive malware detection," in *23rd USENIX Security Symposium (USENIX Security 14)*. San Diego, CA: USENIX Association, Aug. 2014, pp. 287–301. [Online]. Available: https://www.usenix.org/conference/usenixsecurity14/technical-sessions/presentation/kirat

[294] P. Mishra, I. Verma, and S. Gupta, "Kvminspector: Kvm based introspection approach to detect malware in cloud environment," *Journal of Information Security and Applications*, vol. 51, pp. 18–47, 2020.

[295] M. Carbone, D. Zamboni, and W. Lee, "Taming virtualization," *IEEE Security & Privacy*, vol. 6, no. 1, pp. 65–67, 2008.

[296] H. Lee, H. Moon, I. Heo, D. Jang, J. Jang, K. Kim, Y. Paek, and B. B. Kang, "Ki-mon arm: a hardware-assisted event-triggered monitoring platform for mutable kernel object," *IEEE Transactions on Dependable and Secure Computing*, vol. 16, no. 2, pp. 287–300, 2017.

[297] N. L. Petroni Jr, T. Fraser, J. Molina, and W. A. Arbaugh, "Copilot-a coprocessor-based kernel runtime integrity monitor." in *USENIX Security Symposium*. San Diego, USA, 2004, pp. 179–194.

[298] J. Szefer, and R. B. Lee, "Architectural support for hypervisor-secure virtualization," *ACM SIGPLAN Notices*, vol. 47, no. 4, pp. 437–450, 2012.

[299] U. Steinberg, and B. Kauer, "NOVA: a microhypervisor-based secure virtualization architecture," in *5th European Conf. on Computer systems, Paris, France*, 2010, pp. 209–222.

[300] Z. Wang, C. Wu, M. Grace, and X. Jiang, "Isolating commodity hosted hypervisors with hyperlock," in *Proceedings of the 7th ACM European Conference on Computer Systems*, 2012, pp. 127–140.

[301] J. D. Marquez, and M. Castillo, "Performance comparison: Virtual machines and containers running artificial intelligence applications," in *International Conference on Information Technology & Systems*. Springer, 2021, pp. 199–209.

[302] S. Sultan, I. Ahmad, and T. Dimitriou, "Container security: Issues, challenges, and the road ahead," *IEEE Access*, vol. 7, pp. 52 976–52 996, 2019.

[303] B. B. Rad, H. J. Bhatti, and M. Ahmadi, "An introduction to docker and analysis of its performance," *International Journal of Computer Science and Network Security (IJCSNS)*, vol. 17, no. 3, p. 228, 2017.

[304] J. Chelladhurai, P. R. Chelliah, and S. A. Kumar, "Securing docker containers from denial of service (dos) attacks," in *IEEE International Conference on Services Computing (SCC)*. IEEE, 2016, pp. 856–859.

[305] A. Tomar, D. Jeena, P. Mishra, and R. Bisht, "Docker security: A threat model, attack taxonomy and real-time attack scenario of dos," in *10th International Conference on Cloud Computing, Data Science & Engineering (Confluence)*. IEEE, 2020, pp. 150–155.

[306] AWS, *Exploiting AWS ECR and ECS with the Cloud Container Attack Tool (CCAT)*, Sept 2019. [Online]. Available: https://rhinosecuritylabs.com/aws/cloud-container-attack-tool/

[307] R. R. Karn, P. Kudva, H. Huang, S. Suneja, and I. M. Elfadel, "Cryptomining detection in container clouds using system calls and explainable machine learning," *IEEE Transactions on Parallel and Distributed Systems*, vol. 32, no. 3, pp. 674–691, 2020.

[308] M. J. Scheepers, "Virtualization and containerization of application infrastructure: A comparison," in *21st Twente Student Conference on IT*, vol. 21, 2014.

[309] W. Felter, A. Ferreira, R. Rajamony, and J. Rubio, "An updated performance comparison of virtual machines and linux containers," in *2015 IEEE International Symposium on Performance Analysis of Systems and Software (ISPASS)*. IEEE, 2015, pp. 171–172.

[310] K.-T. Seo, H.-S. Hwang, I.-Y. Moon, O.-Y. Kwon, and B.-J. Kim, "Performance comparison analysis of linux container and virtual machine for building cloud," *Advanced Science and Technology Letters*, vol. 66, no. 105-111, p. 2, 2014.

[311] C.-W. Tien, T.-Y. Huang, C.-W. Tien, T.-C. Huang, and S.-Y. Kuo, "Kubanomaly: Anomaly detection for the docker orchestration platform with neural network approaches," *Engineering Reports*, vol. 1, no. 5, p. e12080, 2019.

[312] A. S. Abed, T. C. Clancy, and D. S. Levy, "Applying bag of system calls for anomalous behavior detection of applications in linux containers," in *2015 IEEE Globecom Workshops (GC Wkshps)*. IEEE, 2015, pp. 1–5.

[313] docs.docker, *Install Docker Engine on Ubuntu*, Feb 2021. [Online]. Available: https://docs.docker.com/engine/install/ubuntu/

[314] NSLKDD, *NSLKDD Dataset*, 2009, https://www.unb.ca/cic/datasets/nsl.html.

[315] S. Bhatia, D. Schmidt, G. Mohay, and A. Tickle, "A framework for generating realistic traffic for distributed denial-of-service attacks and flash events," *Computers & Security*, vol. 40, pp. 95–107, 2014.

[316] CAIDA, *CAIDA Dataset*, 2015, https://www.caida.org/data/.

Index

For Product Safety Concerns and Information please contact our EU representative GPSR@taylorandfrancis.com
Taylor & Francis Verlag GmbH, Kaufingerstraße 24, 80331 München, Germany

www.ingramcontent.com/pod-product-compliance
Lightning Source LLC
Chambersburg PA
CBHW060443240326
41598CB00087B/3006

* 9 7 8 1 0 3 2 1 9 0 2 6 6 *